Water Science and Technology Library

Volume 122

The aim of the *Water Science and Technology Library* is to provide a forum for dissemination of the state-of-the-art of topics of current interest in the area of water science and technology. This is accomplished through publication of reference books and monographs, authored or edited. Occasionally also proceedings volumes are accepted for publication in the series. *Water Science and Technology Library* encompasses a wide range of topics dealing with science as well as socio-economic aspects of water, environment, and ecology. Both the water quantity and quality issues are relevant and are embraced by *Water Science and Technology Library*. The emphasis may be on either the scientific content, or techniques of solution, or both. There is increasing emphasis these days on processes and *Water Science and Technology Library* is committed to promoting this emphasis by publishing books emphasizing scientific discussions of physical, chemical, and/or biological aspects of water resources. Likewise, current or emerging solution techniques receive high priority. Interdisciplinary coverage is encouraged. Case studies contributing to our knowledge of water science and technology are also embraced by the series. Innovative ideas and novel techniques are of particular interest.

Comments or suggestions for future volumes are welcomed.

Vijay P. Singh, Department of Biological and Agricultural Engineering & Zachry Department of Civil and Environment Engineering, Texas A&M University, USA Email: vsingh@tamu.edu

All contributions to an edited volume should undergo standard peer review to ensure high scientific quality, while monographs should also be reviewed by at least two experts in the field.

Manuscripts that have undergone successful review should then be prepared according to the Publisher's guidelines manuscripts: https://www.springer.com/gp/authors-editors/book-authors-editors/book-manuscript-guidelines

Elvis Fosso-Kankeu · Ali Al Alili · Hemant Mittal ·
Bhekie Mamba
Editors

Atmospheric Water Harvesting Development and Challenges

Springer

Editors
Elvis Fosso-Kankeu
Department of Mining Engineering, College of Science, Engineering and Technology
University of South Africa, Science Campus
Johannesburg, South Africa

Hemant Mittal
Department of Mechanical Engineering
Khalifa University of Science and Technology
Abu Dhabi, United Arab Emirates

Ali Al Alili
Dubai Electricity and Water Authority (DEWA)
DEWA R&D Center
Dubai, United Arab Emirates

Bhekie Mamba
Institute for Nanotechnology and Water Sustainability
University of South Africa
Johannesburg, South Africa

ISSN 0921-092X ISSN 1872-4663 (electronic)
Water Science and Technology Library
ISBN 978-3-031-21745-6 ISBN 978-3-031-21746-3 (eBook)
https://doi.org/10.1007/978-3-031-21746-3

This Springer imprint is published by the registered company Springer Nature Switzerland AG
The registered company address is: Gewerbestrasse 11, 6330 Cham, Switzerland

Preface

Atmospheric water harvesting basically relies on atmospheric humidity, the principle behind the technology consists of capturing moisture from thin air and the captured moisture is then condensed into liquid water. Atmospheric water harvesting systems mostly make use of atmospheric water generators (AWGs) which capture atmospheric moisture and condense it into freshwater. Many countries in the world, and mostly those experiencing freshwater scarcity, have been using this technology to supplement the shortage in their water resources. It is however important to mention that only little research on atmospheric water harvesters have been conducted to a comprehensive extent in order to facilitate the implementation of this technology in as many countries as possible. Currently, the distribution of atmospheric water harvesting systems remains a huge challenge as only few systems are commercially operating.

The commercial operation of atmospheric water harvesting systems is still limited to few countries; this is mainly due to the low energy efficiency of the system and the inability to effectively operate throughout the various seasons of the year. Researchers have attempted to develop strategies to render the operation of atmospheric water harvesters easier and cost-effective. This book presents ten (10) innovative chapters not published elsewhere, that cover work progress toward such direction, including among others the cooperation of these systems with renewable energy source and the adaptation of the systems to local conditions; the response of the communities in a specific country to such a technology and how its implementation is affected by cultural believe, cost and technological friendliness.

The editors and the publisher are grateful to the reviewers who have contributed to improving the quality of the book through their constructive comments. The editors also thank the publisher for including this book in their series.

This book will be of interest to academic researchers, students, water authorities, professional in relevant industries, government regulatory bodies officers and environmentalists.

Johannesburg, South Africa — Elvis Fosso-Kankeu
Dubai, United Arab Emirates — Ali Al Alili
Abu Dhabi, United Arab Emirates — Hemant Mittal
Johannesburg, South Africa — Bhekie Mamba
December 2022

Contents

About the Editors

Prof. Elvis Fosso-Kankeu has a doctorate degree from the University of Johannesburg in South Africa. He is currently a professor in the Department of Mining Engineering at the University of South Africa. His research focuses on the hydrometallurgical extraction of metal from solid phases, prediction of pollutants dispersion from industrial areas and on the development of effective and sustainable methods for the removal of inorganic and organic pollutants from polluted water. He has published more than 220 papers including journal articles, books, book chapters and conference proceeding papers. He has won several research awards including the NSTF Award (National Science and Technology Forum: largest science, engineering, technology and innovation awards in South Africa and are known as the "Science Oscars" of recent times) Engineering Research Capacity Development, in 2019. Moreover, Prof. Fosso-Kankeu has a H-index of 25 on Research Gate with 2512, an H-index of 27 and more than 2800 citations on Google Scholar.

Prof. Bhekie Mamba is the executive dean of the College of Science, Engineering and Technology, University of South Africa, since January 2017. He previously served as the director of the Nanotechnology and Water Sustainability (NanoWS) Research Unit at the University of South Africa. Prof. Mamba is a visionary and astounding leader and has occupied a number of leadership positions including being a professor and the head in the Department of Applied Chemistry at the University of Johannesburg, the executive dean of the Faculty of Science at the University of Johannesburg, the director of the DST/Mintek Nanotechnology Innovation Centre—Water Research Node—and the director of the Institute of Nanotechnology and Water Research at the University of Johannesburg. Prof. Mamba has published about 7 book chapters, over 250 journal papers, about 12 technical reports and over 50 conference proceedings. He has supervised to completion of over 60 master's and doctoral students who are now either employed or running businesses in SA and other countries in Southern Africa. Besides his established international collaborative research network with other esteemed universities locally and abroad, Prof. Mamba has presented his research work in several local and international conferences. He has reviewed journal articles for at least 20 international journals and has been an external examiner of MSc dissertations and PhD theses from various universities including Wits University, Tshwane University of Technology, University of the Free State, Royal Institute of Science and Technology (Sweden) (Main External Examiner), Rhodes University, University of Western Cape, University of Botswana and University of Stellenbosch. He has vast experience in the area of nanotechnology, polymer chemistry and water treatment technologies. His passion is toward creating sustainable solutions that would ensure that the water resources are maintained and preserved for future generations.

His general research interest involves developing advanced technologies for water treatment, which include nanotechnology and membrane technology. The main interest is the removal of organic micro-pollutants in water and improving the efficiency of conventional technologies in dealing with new emerging pollutants through integrating existing technologies with nanotechnology.

Dr. Hemant Mittal is currently working as a research scientist iat the Chemical Engineering Department, Khalifa University of Science & Technology (KU), Abu Dhabi, United Arab Emirates. He received his PhD degree from Dr. B. R. Ambedkar National Institute of Technology (NIT) Jalandhar, India, in 2013. Dr. Mittal is listed in the Stanford University's list of world's top 2% scientists consecutively three times for the years 2019, 2020 and 2021. His research work focuses on the development of biopolymers-based advanced hydrogels and their nanostructured materials for different industrial applications such as wastewater treatment, atmospheric water harvesting, carbon capture and sea water desalination. He has published over 75 research articles in this field which have been well cited as evident from his H-index of 40.

Dr. Ali Al-Alili is currently working as the vice president of Research & Development at Dubai Electricity & Water Authority (DEWA), UAE. He is in charge of overseeing the entire research program, infrastructure, budgeting and procurement. Prior to joining DEWA, Dr. Al-Alili was an associate professor at Khalifa University, UAE. Dr. Al-Alili received his BSc in Mechanical Engineering from Arizona State University, USA. Then, he joined a combined MSc/PhD program at the University of Maryland, USA, where he received his PhD in Mechanical Engineering.

Chapter 1
Atmospheric Water Generator Technologies

Irfan Majeed Bhat, Ruheena Tabasum, Ghulam Mohd, Kowsar Majid, and Saifullah Lone

1.1 Introduction

Life on Earth is inconceivable without water. Although water cover around 75% of the Earth's surface; nonetheless, a mere 2.8% of it is considered freshwater (Vörösmarty et al. 2000). And the distribution of freshwater is uneven across the world. As a result, the availability of freshwater in some arid regions is critically insufficient. Some four billion people in the world (i.e., two-thirds of the world's population) face low-to-high water stress (Thushantha Harshi Weerasinghe 2013; Bain et al. 2018). With the rapid surge in the global population, finding alternative freshwater resources is imperative to solve the overwhelming water demand. Access to safe drinking water is essential for all, as it has been recognized as an international development priority (United Nations framework for global development priorities, 2030). Given the indiscreet human actions, the existing freshwater sources (including rivers, lakes, and groundwater) are getting depleted at an alarming speed (Thushantha Harshi Weerasinghe 2013). In the following decades, with an improved healthcare system, population expansion and imprudent energy consumption trends could intensify freshwater inadequacy (Vörösmarty et al. 2000; Fuller et al. 2022). On this front, novel approaches have been investigated to avoid the evident future freshwater scarcity menace (Lone et al. 2019).

In this connection, atmospheric water (available regardless of geographical and hydrologic conditions) has been proposed as one of the significant resources of

Note First three authors have contributed equally

I. M. Bhat · R. Tabasum · G. Mohd · K. Majid · S. Lone (✉)
Department of Chemistry, National Institute of Technology (NIT), Jammu and Kashmir, Srinagar 190006, India
e-mail: saifullah.lone@nitsri.net

iDREAM (Interdisciplinary Division for Renewable Energy and Advanced Materials), NIT, Srinagar 190006, India

E. Fosso-Kankeu et al. (eds.), *Atmospheric Water Harvesting Development and Challenges*, Water Science and Technology Library 122,
https://doi.org/10.1007/978-3-031-21746-3_1

freshwater (Rudzewicz and Mata 2007). Earth's atmosphere holds water in the form of water droplets/or vapor (which accounts for up to ~10% of freshwater sources). At any given time, there is seven-fold more water (13 trillion cubic meters) in the Earth's atmosphere than the entire freshwater in rivers and lakes (Muhlfeld et al. 2014). Thus, atmospheric water harvesting (AWH) could be a promising, reliable, and impactful approach to resolving acute water stress in the present and as well as in the future (Mohd et al. 2022; Oki and Kanae 2006; Klemm et al. 2006; Andrews et al. 2011).

Various efforts have been made to explore new sources of freshwaters. Water purification technologies, for instance, filtration (Kadam et al. 2019a), reverse osmosis (Jiao et al. 2020; Fritzmann et al. 2007), multistage flash distillation (Greenlee et al. 2009), and solar water purification (Alsehli et al. 2017; Zhao et al. 2020; Guo et al. 2019; Zhou et al. 2019a; Guo et al. 2020; Zhao et al. 2018), have been thoroughly examined to utilize sea/or wastewater. Nonetheless, such technologies are mainly feasible in coastal areas and are inaccessible for landlocked regions (Zhou et al. 2019b; Tu et al. 2018). It is estimated that the Earth's atmosphere holds ~50,000 km^3 of freshwater in the form of vaporized state (Kalmutzki et al. 2018). In addition, the natural hydrologic cycle enables a sustainable water supply (Zhao et al. 2019). Therefore, AWH is turning out to be an important strategy for water production in landlocked regions of the world. Figure 1.1 illustrates the estimated global water distribution.

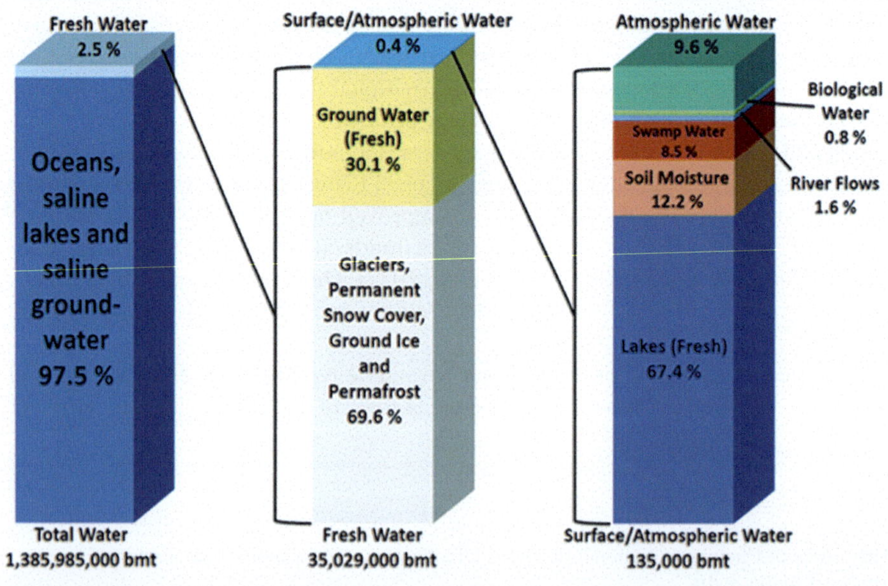

Fig. 1.1 Estimated global water distribution. Copyright©2020-Royal society of chemistry (Peeters et al. 2020)

Despite the high capacity for extractable atmospheric water (Yin et al. 2017; Söz Cal et al. 2020; Parker and Lawrence 2001), only a few AWH systems are commercially operating. Therefore, crafting low-cost atmospheric water fishing technologies and distributing them evenly at an affordable price is the way forward. In general, any viable AWH technology must satisfy four key criteria. It should be efficient, cheap, stable, and cyclic. Currently, none of the existing commercial atmospheric water harvesting generators (AWHGs) satisfies all four parameters. It is mainly due to the energy inefficiency of the process, which could be rationalized from a thermodynamics point of view. Artificial rain, fog, and dew water collection are the various categories of water harvesting from the atmosphere. Different technologies, problems, and perspectives of AWHGs briefly are summarized in Table 1.1. However, based on the harvesting mechanism, this book chapter will purely focus on two major AWHGs technologies: sorption and condensing.

1.2 AWH by Moisture Sorption Technology

Moisture sorption technology captures water from thin air via absorption\or adsorption process. As shown in Fig. 1.2, the water molecules in air cling to the hierarchical surface of harvesters through physical/ or chemical processes (Kadam et al. 2019b; LaPotin et al. 2019). Among the two; absorption is a bulk phenomenon, it involves the diffusion of gas/liquid molecules into liquid/solid materials, which consequently changes the structure and volume of the absorbents. Further, stoichiometry and concentration of the reactants manipulate the chemical absorption, while the osmotic effect guides the physical absorption (Kadam et al. 2019b; LaPotin et al. 2019; Agam and Berliner 2006; Wang et al. 2005). Contrary, adsorption is a surface feature that ensures gas/or liquid molecules adhere to a solid surface via chemical (chemisorption) or physical (physisorption) interactions, Fig. 1.2b (Gido et al. 2016; Janchen et al. 2004). Adsorption is an intrinsic material property; for instance, in chemical adsorption, binding sites are required for sorbent to chemically (via hydrogen bonding, coordination effect, and electrostatic interactions) bridge with adsorbing molecules (i.e., water). A high energy barrier (80–400 kJ/mol) needed in this case is irreversible without an external energy supply (Gido et al. 2016; Janchen et al. 2004; Ng et al. 2001).

To craft superior moisture harvesters from various smart materials; high sorption ability, low regeneration energy demand, fast sorption/desorption, large surface area, and long-term cycling stability are the essential properties to be taken into consideration (Jin et al. 2017). High sorption ability could be improved by selecting sorption materials with water affinity, large surface area, and high porosity to enhance moisture capturing for vapor liquefaction, and deliver harvested water. Whereas low-water energy requirements could be fulfilled by tuning the sorption characteristic of the harvesting material by impregnating functional materials (for instance; phase change

Table 1.1 A summary of various AWH methods with problems and perspectives

Methods	Features	Capacity	Problems	Alternatives	Challenges	Applications
Cloud seeding	Hygroscopic seeding	NA	Considerable skepticism	Sky river proposed by Wang	Hard to predict cloud migration	Weather modification
Fog seeding	Hygroscopic salt particles	NA	Hard to predict fog occurrence	Dew collection by using PCM	PCM: sorption solidification	Microclimate management
Fog mesh	Easy to construct	1.5–12 kg/day/m^2	Limited to locations	Nano-engineering mesh surface	Low collection efficiency	Mount by the sea
Massive dew collector	Passive way	Very little output	Thin valid layer	PCM dew collector	Lack of high e* materials	Water-saving agriculture
Radiative dew collector	Passive way	0.3-0.6 kg/day/m^2	Large heat loss	High-performance emitter	Metamaterials	Low-T water production
Electric chiller Sorption chiller	High output compact	0.25 Wh/g (electricity)	Great latent load	DEHP	Durability and reliability	Portable water production
Sorption-based AWG (solar distiller)	Ambient cooling	1.0–2.5 kg/day/m^2	High T_{cond}	Water-cooling condenser	Heat transfer enhancement	Scalable water supply
Sorption-based AWG (solar air heating)	Active condenser	2.0 kWh/kg (heat)	Large heat loss	Solar water-heating sorber	Pump	Scalable water supply
Sorption-based AWG (sandwich plate)	Air cooling	1.2 kg/day/m^2	Thickness of desiccant layer	High k desiccant	Desiccant	Electronic cooling
TEC dew collector (solar PV driven)	Small and compact	COP < 0.1	Low energy efficiency	Desiccant-enhanced heat sink	Proper application	Meet minimum water load

(continued)

Table 1.1 (continued)

Methods	Features	Capacity	Problems	Alternatives	Challenges	Applications
Integrated system	By-product	Depends on the latent load	Cooling dependence	NA	NA	Offsetting water use of A/C

Copyright © 2018, Cell press reviews

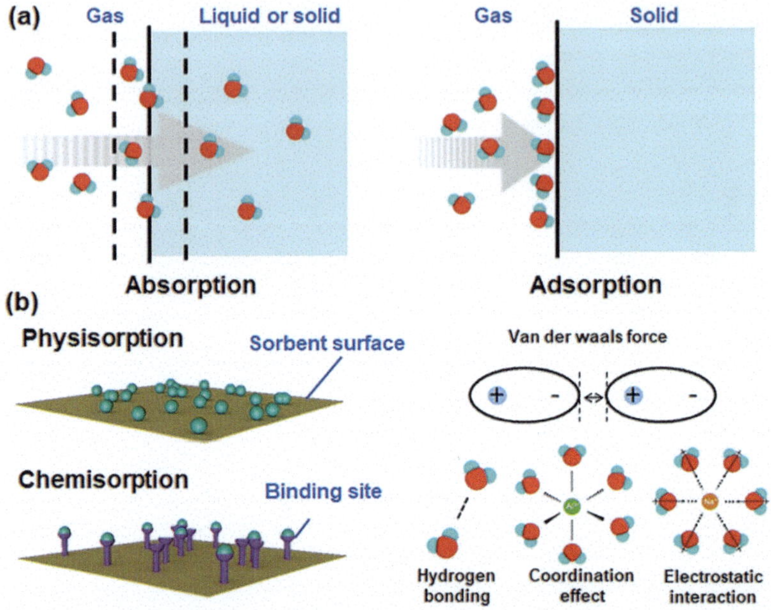

Fig. 1.2 Moisture sorption mechanisms. **a** Schematic of absorption (left) and adsorption (right) processes. **b** Schematic physisorption (top left, top title: atmospheric water generator technologies right) and chemisorption (bottom left, bottom right). Copyright © 2020, American Chemical Society (Zhou et al. 2020)

materials (PCMs)), which release energy during the transition process. And, for efficient sorption harvesters, quick sorption/or desorption over a sustainable period of time (without potential performance decay) is equally critical (Refer to Fig. 1.3).

It is noteworthy to mention that sorption-based AWH technology is more feasible and energy efficient as it uses sorbents and low-grade heat to capture moisture from thin air. The process starts with the capturing of atmospheric moisture, which is followed by condensation. In particular, some sorbents completely rely upon sunlight to harvest water, especially in areas where the relative humidity is low. For superior efficiency at relatively lower humidity, the sorbent material should have a higher affinity for water. However, this strong adhesion would restrict the desorption rate. Therefore, balanced interaction between the sorption material and water is of vital significance to the harvesting process. The moisture sorbent with a high affinity towards water molecules captures moisture from the atmosphere. And, when heated, the concentrated vapor will be released from the sorbent, and then condensed and collected as liquid water. Once the water is released, the sorbents get ready for the next cycle. Besides, the sorbent material plays a vital role in AWH, while the system-level design related to heat and mass transport also holds the key to improving the overall performance of the water harvesting process. Important characteristics of sorbent material (including wide applicable range, fast sorption rate, easy regeneration, high

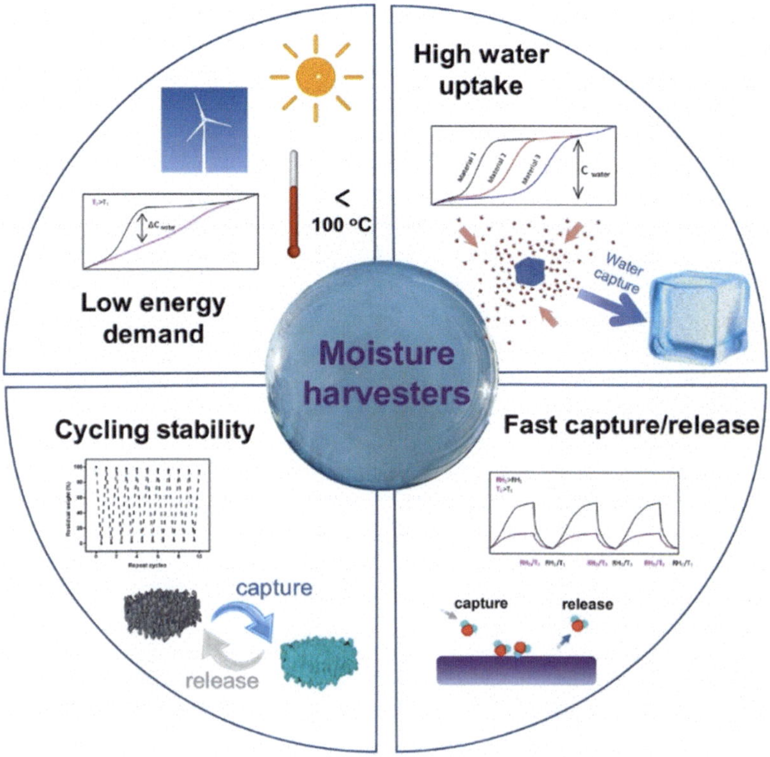

Fig. 1.3 Properties of moisture harvesters. Copyright © 2020, ACS (Zhou et al. 2020; Jin et al. 2017)

water absorbing capacity, and high relative humidity range) are also critically important for are needed for superior efficiency of the harvesting system. Depending upon the properties and structure of the sorbent material, sorption mechanisms include surface adsorption, micropore filling, capillary condensation, and bulk absorption.

1.3 AWH by Vapor Condensing Technology

Water vapor condensing is one of the most common AWH technology (Wahlgren 2001). It can be broadly categorized into fog water collection and dew water collection (Klemm et al. 2001). Fog fishing is a formidable approach to supplying sustainable potable water in arid regions (Bhushan 2020). In contrast, dew catching is performed by passing humid air over a cooled surface. And vapors are condensed liquid water when the surface temperature is below the dew point temperature (Fig. 1.4).

In the case of vapor condensing technology, AWHG is placed perpendicular to the moving air. It traps fog in the form of tiny beads of water droplets on a solid

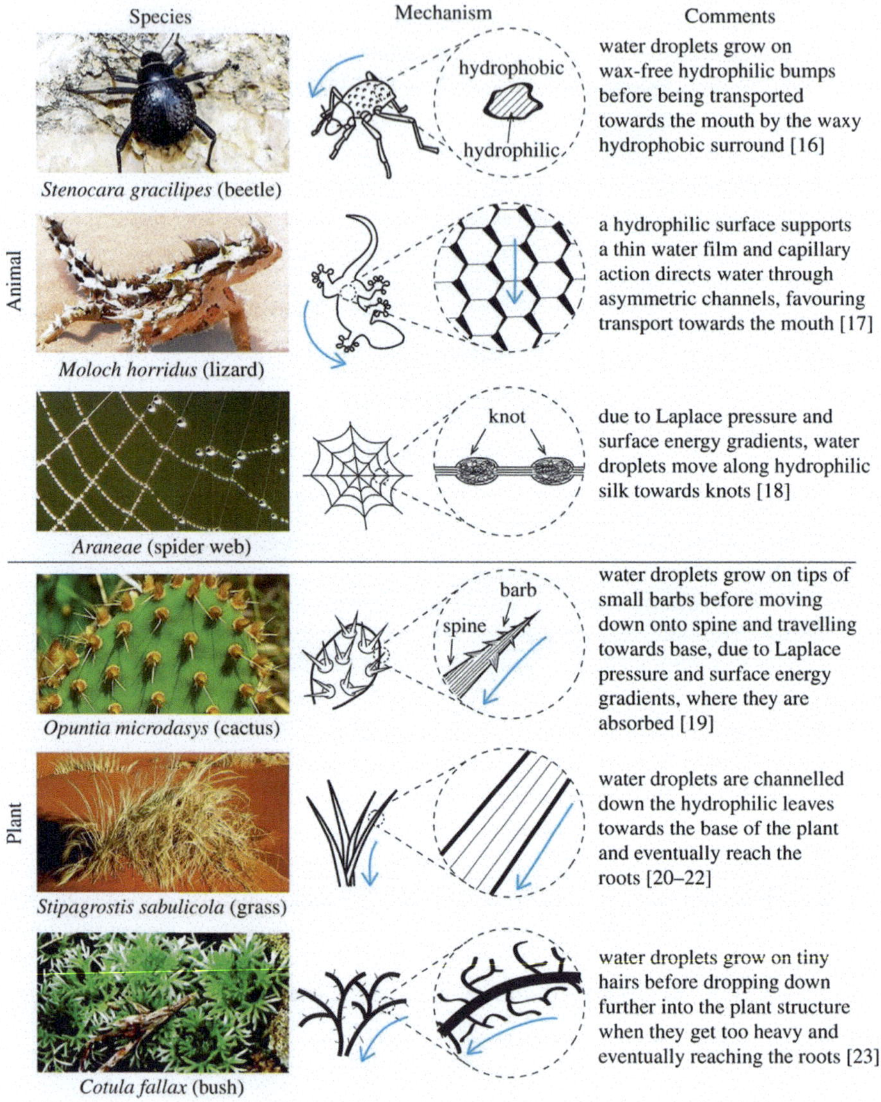

Fig. 1.4 Hierarchical designs found in nature for AWH. Copyright © 2016, Royal Society (Bhushan 2020)

surface (Yin et al. 2017). With time, the water beads get enlarged and coalesced by continuous trapping of the air moisture (Zhang et al. 2017). Finally, the fused water droplets fall under gravity and are collected and stored (Adera et al. 2020). The entire process is repeated for an extended period (Ju et al. 2013). The cyclic method comprises four main steps (nucleation, droplet fusion/or coalescence, droplet mobility/transportation, and water collection) (Fig. 1.5a, b) (Vuollekoski et al. 2014;

Oki and Kanae 2006). The individual steps impact the overall efficiency of AWH and are critically dependent on the surface structure/or properties of the AWGs (Andrews et al. 2011). And the wetting properties of the substrate manipulate the droplet nucleation, growth, and mobility.

The interaction of the water droplet with the solid surface is mainly of the Van der Waals type (Azioune et al. 2002). Therefore, due to substantial polar interaction, the hydrophilic polar substrates can serve better for capturing atmospheric water (Velzenberger et al. 2009). As the small beads of water are absorbed on the substrate, the beads fuse to form larger droplets, which get detached from the nucleating site under various factors, but mainly under the gravity impact. Thus, crafting superior efficiency of AWHGs, the droplet, once formed, should be dislodged as quickly as possible (Kim et al. 2018). This is required for the transportation of the harvested droplets as well as for the cyclic harvesting ability of the system (Li et al. 2018). However, the presence of hydrophilic character in the substrate, initially responsible and favorable for the nucleation step, has a reverse unfavorable effect on the transportation step (Fig. 1.6).

The droplet, once formed, gets pinned on the substrate leading to the blocking of the nucleating site and thereby reducing the efficiency of the harvesting system (Yao

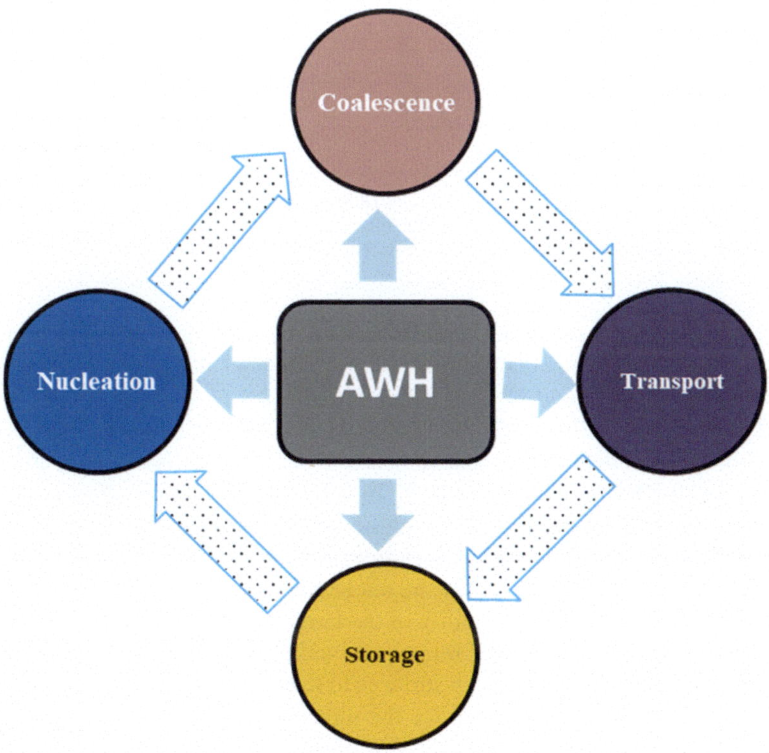

Fig. 1.5 Schematic illustration of the steps involved in the AWH

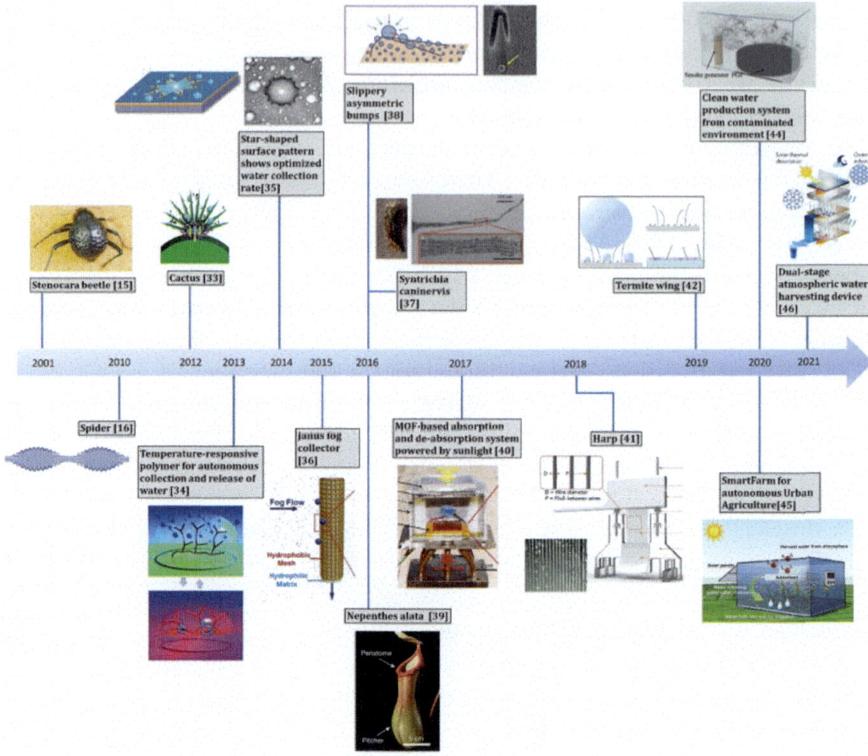

Fig. 1.6 Timeline of the milestones of the bioinspired water harvesting behavior of Stenocara beetle, spider silk, cactus, *Nepenthes alata*, *Syntrichia caninervis*, Termite wing, temperature-responsive collection, and release of water, MOF-based absorption and desorption system powered by sunlight, etc. Copyright © 2020, Royal Society (Bhushan 2020; Cao et al. 2012)

et al. 2010). Due to the substrate's hydrophilic character, this droplet's pinning and blocking effect not only hinders the transportation step but also masks the nucleating sites, which otherwise could be used for another cycle of harvesting (Fathieh et al. 2018). Therefore, no further nucleation takes place. As a result, the cyclic harvesting ability of the system deteriorates and vanishes. Currently, this is the main challenge of most AWHGs (Kandilian et al. 2011). To solve the problem of droplet pinning, a hybrid surface containing both hydrophilic and hydrophobic surfaces is created (Jeyachandran et al. 2009). These surfaces are combined and well-arranged so that the hydrophilic surface is responsible for the nucleation step. The hydrophobic surface helps in the transportation and mobility of the harvested water droplet (Zhang et al. 2015). The combination of hydrophilic spots and slippery areas can also create a wettability gradient in the harvesting surfaces to reduce the droplet pinning problem (Ray et al. 2016). The driving force for the directional liquid transport wettability gradient is necessary. Inspired by nature (for instance, desert beetle, the spider web, the cactus plant, and other such systems), superior AWHGs are fabricated (Ju et al.

2012; Zheng et al. 2010). Efforts have been made to improve the water harvesting efficiency of AWHGs. For directional liquid transport, uniformly rough topography with specific hierarchical structures is significantly essential. The surface roughness and the curvature gradient also transport the condensate water droplet (Dai et al. 2018). The pressure difference called Laplace pressure (ΔP) is responsible for taking the droplet from higher curvature to a lower curvature surface (Menger 1979).

Acknowledgements This work was supported by SERB (Science and Technology Research Board—a statutory body under the Department of Science and Technology, Government of India); under the Research grant of ***Ramanujan Fellow Award (File number: SB/S2/RJN-013/2018).***

Conflict of Interest The authors report no conflict of interest.

References

Adera S, Alvarenga J, Shneidman AV, Zhang CT, Davitt A, Aizenberg J (2020) Depletion of lubricant from nanostructured oil infused surfaces by pendant condensate droplets. ACS Nano 14(7):8024–8035

Agam N, Berliner PR (2006) Dew formation and water vapor adsorption in semi-arid environments. J Arid Environ 65:572–590

Alsehli M, Choi JK, Aljuhan MA (2017) Novel design for solar-powered multistage flash desalination. Sol Energy 153:348–359

Andrews H, Eccles E, Schofield W, Badyal J (2011) Three-dimensional hierarchical structures for fog harvesting. Langmuir 27:3798–3802

Azioune A, Chehimi MM, Miksa B, Basinska T, Slomkowski S (2002) Hydrophobic protein−polypyrrole interactions: the role of van der Waals and Lewis acid−base forces as determined by contact angle measurements. Langmuir 18(4):1150–1156

Bain R, Johnston R, Mitis F, Chatterley C, Slaymaker T (2018) Development goal baselines for household drinking water, sanitation and hygiene services. Water 10:1711

Bhushan B (2020) Design of water harvesting towers and projections for water collection from fog and condensation. Phil Trans R Soc A 378:20190440

Cao Y, Chen Y, Sun X, Zhang Z, Mu T (2012) Water sorption in ionic liquids: kinetics mechanisms and hydrophilicity. Phys Chem 14:12252–12262

Dai X, Sun N, Nielsen SO, Stogin BB, Wang J, Yang S, Wong T S (2018) Hydrophilic directional slippery rough surfaces for water harvesting. Sci Adv 4:eaaq0919

Fathieh F, Kalmutzki MJ, Kapustin EA, Waller PJ, Yang J, Yaghi OM (2018) Practical water production from desert air. Sci Adv 4:eaat3198

Fritzmann C, Löwenberg J, Wintgens T, Melin T (2007) State of the-art of reverse osmosis desalination 216:1–76

Fuller R, Landrigan PJ, Balakrishnan K, Bathan G, Reilly SB, Brauer M (2022) Pollution and health: a progress update. Lancet Planetry Health 6(6):E535–E547

Gido B, Friedler E, Broday DM (2016) Assessment of atmospheric moisture harvesting by direct cooling. Atmos Res 182:156–162

Greenlee LF, Lawler DF, Freeman BD, Marrot B, Moulin P (2009) Reverse osmosis desalination: water sources: technology: and today's challenges. Water Res 43:2317–2348

Guo Y, Zhou X, Zhao F, Bae J, Rosenberger B, Yu G (2019) Synergistic energy nanoconfinement and water activation in hydrogels for efficient solar water desalination. ACS Nano 13:7913–7919

Guo Y, Lu H, Zhao F, Zhou X, Shi W, Yu G (2020) Biomass-derived hybrid hydrogel evaporators for cost-effective solar water purification. Adv Mater 32:1907061

Janchen J, Ackermann D, Stach H, Brosicke W (2004) Studies of the water adsorption on zeolites and modified mesoporous materials for seasonal storage of solar heat. Sol Energy 76:339–344

Jeyachandran YL, Mielczarski E, Rai B, Mielczarski JA (2009) Quantitative and qualitative evaluation of adsorption/desorption of bovine serum albumin on hydrophilic and hydrophobic surfaces. Langmuir 25(19):11614–11620

Jiao M, Yao Y, Chen C, Jiang B, Pastel G, Lin Z, Wu Q, Cui M, He S, Hu L (2020) Highly efficient water treatment via a wood-based and reusable filter. ACS Mater Lett 2:430–437

Jin Y, Zhang L, Wang P (2017) Atmospheric water harvesting: role of surface wettability and edge effect. Global Chall 1:1700019

Ju J, Bai H, Zheng Y, Zhao T, Fang R, Jiang L (2012) Multi-structural and multifunctional integrated fog collection system in cactus. Nat Commun 3:1247–2253

Ju J, Xiao K, Yao X, Bai H, Jiang L (2013) Bioinspired conical copper wire with gradient wettability for continuous and efficient fog collection. Adv Mater 25:5937–5942

Kadam AA, Lone S, Shinde S, Yang J, Saratale RG, Saratale GD, Sung J-S, Young D, Kim, G Ghodake, (2019a) Treatment of hazardous engineered nanomaterials by supermagnetized α-cellulose fibers of renewable paper-waste origin. ACS Sustain Chem Eng 7(6):5764–5775

Kadam A, Ganesh R, Surendra S, Jiwook S, Kyojung Y, Bhupendra H, Ganesh M, Saratale D, Lone S, Kim DY, Sung JS, Ghodake G (2019b) Adsorptive remediation of cobalt oxide nanoparticles by magnetized α-cellulose fibers from waste paper biomass. Biores Technol 273:386–393

Kalmutzki MJ, Diercks CS, Yaghi OM (2018) Metal-organic frameworks for water harvesting from air. Adv Mater 30:1704304

Kandilian R, Navid A, Pilon L (2011) The pyroelectric energy harvesting capabilities of PMN–PT near the morphotropic phase boundary. Smart Mater Struct 20(20):055020

Kim H, Rao SR, Kapustin EA, Zhao L, Yang S, Yaghi OM, Wang EN (2018) Adsorption-based atmospheric water harvesting device for arid climates. Nat Commun 10(7):018–03162

Klemm O, Schemenauer RS, Lummerich A, Cereceda P, Marzol V, Corell D, Van Heerden J, Reinhard D, Gherezghiher T, Olivier J (2001) Fog as a fresh-water resource: overview and perspectives. Ambio 41:221–234

Klemm O, Schemenauer RS, Lummerich A, Cereceda P, Marzol V, Corell D, Van Heerden J, Reinhard D, Gherezghiher T, Olivier J (2006) Fog as a fresh-water resource: overview and perspectives. Ambio X 41:221–234

LaPotin A, Kim H, Rao SR, Wang EN (2019) Adsorption-based atmospheric water harvesting: impact of material and component properties on system-level performance. Acc Chem Res 52:1588–1597

Li R, Shi Y, Alsaedi M, Wu M, Shi L, Wang P (2018) Hybrid hydrogel with high water vapor harvesting capacity for deployable solar-driven atmospheric water generator. Environ Sci Technol 52:11367–11377

Lone S, Yoon DH, Lee H, Cheong IW (2019) Gelatin–chitosan hydrogel particles for efficient removal of Hg (II) from wastewater. 5:83–90

Menger FM (1979) Laplace pressure inside micelles. J Phys Chem 83(38):893

Mohd G, Majid K, Lone S (2022) Multiscale Janus surface structure of Trifolium leaf with atmospheric water harvesting and dual wettability features. ACS Appl Mater Interfaces 14(3):4690–4698

Muhlfeld CC, Kovach RP, Jones LA, Al-Chokhachy R, Boyer MC, Leary RF, Lowe WH, Luikart G, Allendorf FW (2014) Invasive hybridization in a threatened species is accelerated by climate change. Nat Clim Chang 4:620–624

Ng K, Chua H, Chung C, Loke C, Kashiwagi T, Akisawa A, Saha BB (2001) Experimental investigation of the silica gel water adsorption isotherm characteristics. Appl Therm Eng 21:1631–1642

Oki T, Kanae S (2006) Global hydrological cycles and world water resources. Science 313:1068–1072

Parker AR, Lawrence CR (2001) Water capture by a desert beetle. Nature 414:33
Peeters R, Vanderschaeghe H, Ronge J, Johan A (2020) Martens Energy performance and climate dependency of technologies for fresh water production from atmospheric water vapour. Environ Sci Water Res Technol 6:2016–2034
Ray S, Ghosh B, Bardhan S, Bhattacharyya B (2016) Studies on the impact of energy quality on human development index. Renew Energy 92:117–126
Rudzewicz ZW, Mata LJ (2007) Freshwater resources and their management. Cambridge University Press, pp 173–210
Söz Cal K, Trosien S, Biesalski MA (2020) Critical review and comparative study. ACS Mater Lett 2:336–357
Thushantha Harshi Weerasinghe WM (2013) Designing a dynamically integrated water management scheme as an adaptation strategy for global change-induced water stress. Earth Syst Sci 1614–1199
Tu Y, Wang R, Zhang Y, Wang J (2018) Progress and expectation of atmospheric water harvesting. Joule 2:1452–1475
Velzenberger E, ElKirat K, Gilbert LM, Nagel D, Pezron I (2009) Characterization of biomaterials polar interactions in physiological conditions using liquid–liquid contact angle measurements: relation to fibronectin adsorption. Colloids Surf 68(2):238–244
Vörösmarty CJ, Green P, Salisbury J, Lammers RB (2000) Global water resources: vulnerability from climate change and population growth. Science 289:284–288
Vuollekoski H, Vogt M, Sinclair VA, Duplissy J, Jarvinen H, Kyro EM, Makkonen R, Petaja T, Prisle NL, Raisanen P (2014) Estimates of global dew collection potential. Hydrol Earth Syst 11:9519–9549
Wahlgren RV (2001) Atmospheric water vapor processor designs for potable water production. Water Res 35:1–22
Wang D, Xia Z, Wu J, Wang R, Zhai H, Dou W (2005) Study of a novel silica gel water adsorption chiller design and performance prediction. Int J Refrig 28:1073–1083
Yao X, Song Y, Jiang L (2010) Applications of bio-inspired special wettable surfaces. Adv Mater 23(6):719–734
Yin K, Du H, Dong X, Wang C, Duan JA, He JA (2017) Simple way to achieve bioinspired hybrid wettability surface with micro/nano patterns for efficient fog collection. Nanoscale 9:14620–14626
Zhang L, Wu J, Hedhili MN, Yang X, Wang P (2015) Inkjet printing for direct micropatterning of a superhydrophobic surface toward biomimetic fog harvesting surfaces. J Mater Chem A 3(25):2844–2852
Zhang H, Yoshino H, Hasegawa K, Liu J, Zhang W, Xuan H (2017) Practical moisture buffering effect of three hygroscopic materials in real world conditions. Energy Build 139:214–223
Zhao F, Zhou X, Shi Y, Qian X, Alexander M, Zhao X, Mendez S, Yang R, Qu L, Yu G (2018) Highly efficient solar vapor generation via hierarchically nanostructured gels. Nat Nano Technol 13:489–495
Zhao F, Zhou X, Liu Y, Shi Y, Dai Y, Yu G (2019) Super moisture-absorbent gels for all weather atmospheric water harvesting. Adv Mater 31:1806446
Zhao F, Guo Y, Zhou X, Shi W, Yu G (2020) Materials for solar-powered water evaporation. Nat Rev Mater 5:388–401
Zheng Y, Bai H, Huang Z, Tian X, Nie FQ, Zhao Y, Zhai J, Jiang L (2010) Directional water collection on wetted spider silk. Nature 463(31):640–643
Zhou X, Guo Y, Zhao F, Yu G (2019a) Hydrogels as an emerging material platform for solar water purification. Acc Chem Res 52:3244–3253
Zhou X, Zhao F, Guo Y, Rosenberger B, Yu G (2019b) Architecting highly hydratable polymer networks to tune the water state for solar water purification. Sci Adv 5:eaaw5484
Zhou X, LuFei H, Yu ZG (2020) Atmospheric water harvesting: a review of material and structural designs. ACS Mater Lett 2(7):671–684

Chapter 2
Outdoor Testing of Double Slope Condensation Surface for Extraction of Water from Air

Karim H. Awad, Mohamed M. Awad, and Ahmed M. Hamed

2.1 Introduction

The World Health Organization (WHO) claimed that around 20% of the world's population live in regions where there is insufficient water to meet their needs (United Nations World Water Assessment Program 2012). In the present day, water requirements for drinking, production of food, and agriculture are viewed as one of the most urgent problems that face the world. This is a direct effect of the shortage and lack of freshwater sources, as well as the growing salinity of groundwater. With the global population growing by 80 million people every year, a third of the world's population would likely confront water scarcity by 2025. This threatening water crisis is connected to food production because agriculture uses 70% of all freshwater, and acquiring irrigation water in dry regions using traditional methods has significant environmental consequences. Many countries have a significant amount of solar energy potential. In arid regions, this high solar energy potential can be best used to generate water for irrigation. This high solar energy potential is best used to generate water for irrigation in arid areas.

K. H. Awad · M. M. Awad · A. M. Hamed (✉)
Mechanical Power Engineering Department, Faculty of Engineering, Mansoura University, Mansoura 35516, Egypt
e-mail: amhamed@mans.edu.eg

M. M. Awad
e-mail: m_m_awad@mans.edu.eg

K. H. Awad
High Institute of Engineering and Technology, New Damietta, Damietta, Egypt

E. Fosso-Kankeu et al. (eds.), *Atmospheric Water Harvesting Development and Challenges*, Water Science and Technology Library 122,
https://doi.org/10.1007/978-3-031-21746-3_2

2.2 Providing Fresh Water to Arid Regions

The suitable solution for solving the problem of providing fresh water in these arid zones relies upon the climatic, geographical, extent of the ability to create the infrastructure and quantity of water required. This issue can be solved using one of the following techniques (Hamed 2000):

1. Water transport from other locations;
2. saline water desalination (ground and underground);
3. Harvesting water from atmospheric air.

Water transportation in these places is typically quite expensive and has a significant initial cost. Also, Desalination relies on the presence of saline water supplies, which are uncommon in these regions. Some countries close to the sea rely on desalinating water as a major source of water, but it has height cost. Atmospheric air is a tremendous and inexhaustible water reservoir. This limitless source of water is available anywhere on the planet's surface. The amount of water in the atmosphere is estimated to be 14,000 km^3, whereas fresh water on the planet is only about 1200 km^3 (Obrezkova 1988).

2.3 Extraction of Water from Atmospheric Air

There are three common techniques to generate water from atmospheric air. First, lowering the temperature of moist air below the dew point. Second, wet accumulation from the fog. Third, using a solid or liquid desiccant to absorb water vapor from moist air, and then recovering the extracted water by heating the desiccant and condensing the evaporated water. The last method can be correlated with solar radiation as a heating source. The season has a height effect on harvesting water from the air, where the average amount of water vapor in the air and solar radiation intensity varies from one season to another all over the world as observed by National Aeronautics and Space Administration (NASA) Aqua satellite in Fig. 2.1 (NASA Earth Observatory 2022).

2.4 Commercial Applications of Extracting Water from Atmospheric Air

Inventors and investors have developed unconventional methods to generate water. They produced water harvesting systems from the air from laboratory research to commercial applications. It can be utilized in dry areas for industrial, commercial and residential applications. In 2015, the market for air water generators from cooling and condensation was estimated at over USD 800 million. Advancements in technology to

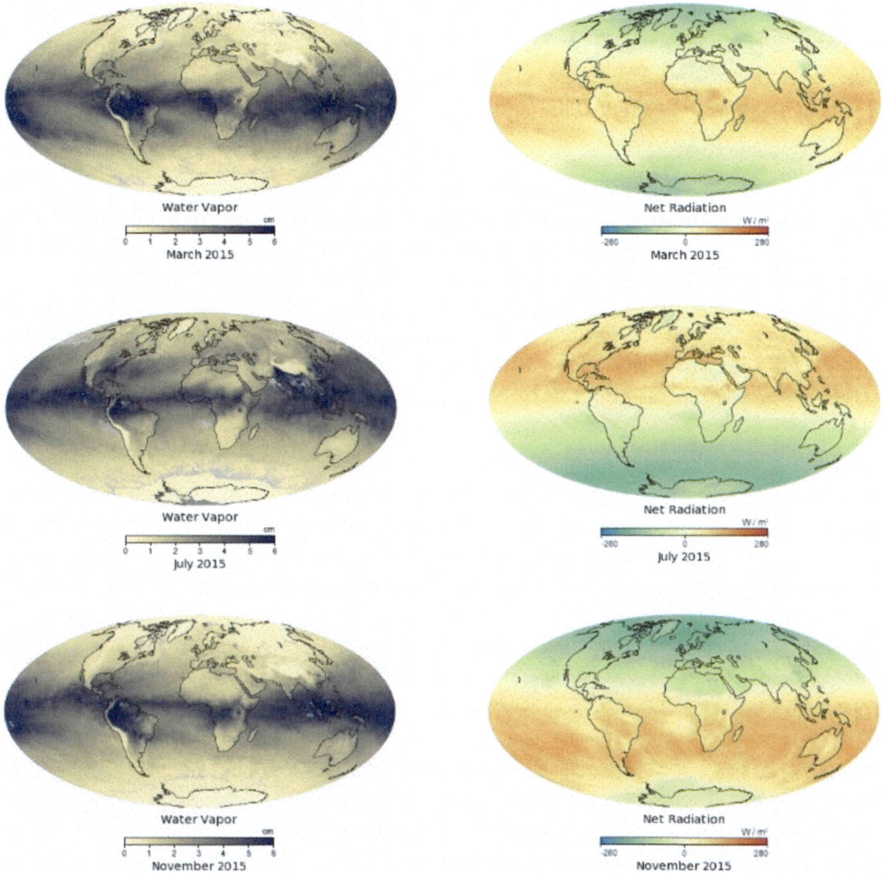

Fig. 2.1 Maps of global water vapor in air and net radiation (NASA Earth Observatory 2022)

minimize equipment and electricity costs will support industry growth during forecast periods. The market for atmospheric water generators (AWGs) is expected to reach \$9.3 billion by 2022, with a compound annual growth rate (CAGR) of 37.4% from 2015 to 2022, as indicated by a new review based report by Global Market Insights (Market Research Reports, Consulting 2022). In the future, the global atmospheric water generator market share will be driven by declining freshwater paired with increased foundation investment.

Watair is one of the companies which is engaged in using a technique of extracting water from the air (WatAir 2022). Since 2001, It has developed and brought to the manufacturing stage items for harvesting liquid water from atmospheric air and treating it to make it safe to drink. It has produced modules for home/office and commercial/industrial responses. It has introduced its home/office product in 2003 under the name of AirJuicer as shown in Fig. 2.2. Its production up to 24 L of water per day. Now, their atmospheric water generators make freshwater from atmospheric air from 15 L per day to 10,000 L per day. They introduced three models of home applications (PW HR-15L, PW HR-30 L, and PW HR-60L), and seven models of industrial applications (PW HR-100 L, PW HR-250 L, PW HR-500 L, PW HR-1000 L, PW HR-3000 L, PW HR-5000 L, and PW HR-10,000 L). These generators operate at temperatures of 15–40 °C and 35–95% humidity, but the target quantity for each generator is generated at 30 °C and 80% RH.

American company SKYWELL has developed two modules of water generator from air (Product 2022). The first module calls SKYWELL5, which is shown in Fig. 2.3 and it can make up to five gallons in a day. The other module calls SKYWELL100 can make up to 100 gallons in a day.

A water production company Water-gen has developed modules for home, commercial and mobility generators (Watergen 2022). They introduced three modules of commercial generators (GEN-M1, GEN-M PRO, and GEN-L) which

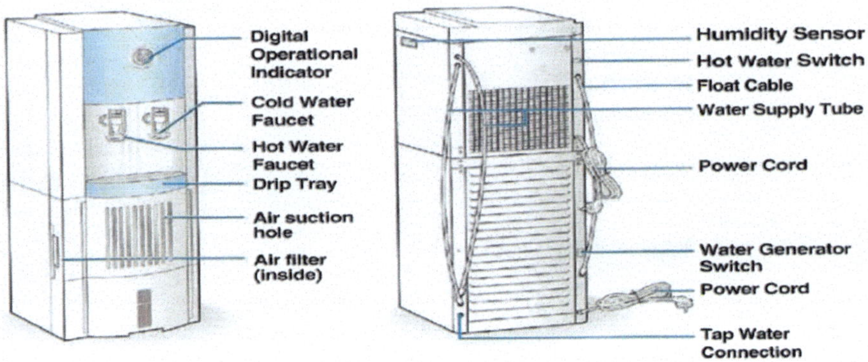

Fig. 2.2 Schematic of AirJuicer unit (WatAir 2022)

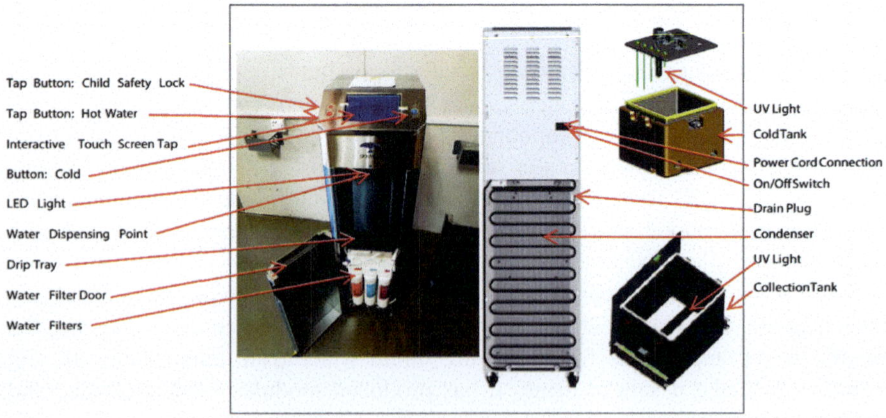

Fig. 2.3 Photograph of SKYWELL 5 parts (Product 2022)

Provides up to 220,900, and 6000 L of clean fresh water every day, respectively. Their home generator Makes up to 30 L. Their portable generators called "Watergen Mobile Box" that produces up to 20 L per day of drinking water. Recently, they introduced a new module of portable generators called "Watergen ON Board" that produces up to 50 L per day of drinking water. Figure 2.4 shows the technology of their generators.

WaterMaker Company is a social entrepreneur in India who provides technology for clean and safe drinking water. They design and manufacture a large number of atmospheric water generators (WaterMaker India 2022). These machines may provide anywhere from 25 to 2500 L of pure drinking water per day. They introduced home generator called "magic" which makes 25 L of fresh water. They introduced five module of industrial generator which are already in use in India and many countries around the world.

Maithri Aquatech company was established with the mission to provide water on a sustainable basis wherever required in India (Our Story 2022). The Indian Institute of Chemical Technology (IICT), which is associated with the Council of Scientific and Industrial Research (CSIR), has partnered with Maithri Aquatech. The company has developed a technology to produce potable drinkable water from atmospheric air. Its innovative solution called MEGHDOOT with four series. These machines may generate anywhere from 60 to 2000 L of pure drinking water per day.

Most commercial applications to extract water from air humidity are confined to the cooling process. However, there is a need to study the system in conditions similar to that of the arid zones with small modification of the ground layer surface using desiccant. For this purpose, the sand bed impregnated with solution of a Calcium Chloride is used as absorber and solar radiation as a heat resource. A triangular prism

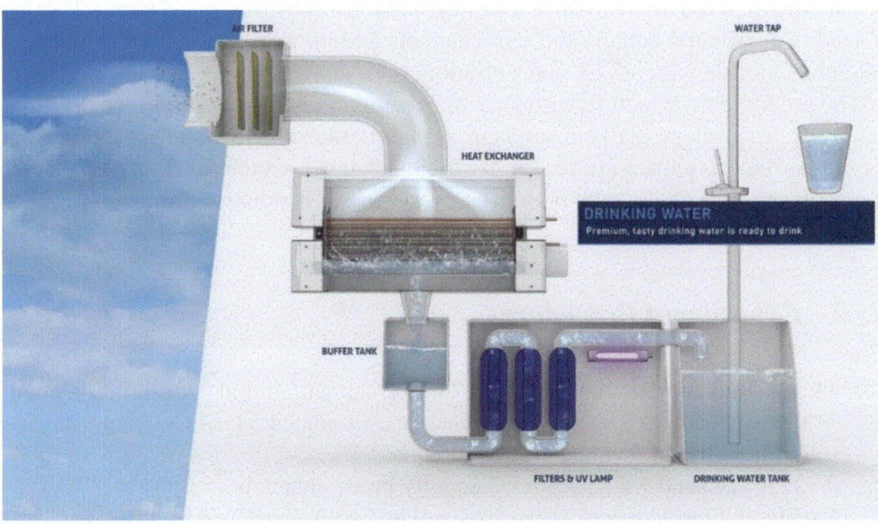

Fig. 2.4 Technology of water-gen's water generator machine (Watergen 2022)

surface is used as transparent and condensation surface which cover the bed at day-time. The purpose of this experimental and theoretical research is to investigate the impact of various operating factors (solar radiation intensity, ambient temperature and bed configurations) on the performance of the water extraction system.

2.5 Experimental Study of the Model

The experiments are carried out on the roof of the Combustion Laboratory in open air, Faculty of Engineering, Mansoura University, Department of Mechanical Power Engineering. The study was performed in August, September and October at climatic conditions of Mansoura, Egypt (31.04 °N and 31.23 °E).

2.5.1 Description of Apparatus

In the present study, a double sloped condensation surface covering a layer of the ground surface will be applied for regeneration and condensation of water. The ground layer, which is impregnated with Calcium Chloride ($CaCl_2$) solution, will function as an absorber. For this purpose, an apparatus consists of two parts is designed and fabricated.

The first part, which functions as absorber, has a triangular prism (200 cm × 37 cm × 36 cm); the bottom is covered with fiber plate (50 cm × 200 cm). Channel (11 cm × 200 cm) is fixed at the top of a triangular prism and eight channels (5 cm × 200 cm) are fixed at the main sides of the prism as shown in Fig. 2.5. A 0.005 m^3 of sand is distributed equally through channels and impregnated with a solution of Calcium Chloride with a 37% concentration.

The second part, which functions as transparent and condensation surface, has a triangular prism shape (210 cm × 65 cm × 51 cm), the sides are made of transparent acrylics surface as shown in Fig. 2.6. A sloping channel, which function as collector of condensate, are fixed at the bottom of every side of condensation surface.

2.5.2 Experiment Procedures

1. 4 kg of solid commercial Calcium Chloride salt with initial concentration about 74% has been mixed with 4 kg of water and produced 5 L of Calcium Chloride solution with initial concentration of about 37%.
2. 0.005 m^3 of sand was distributed equally in the channels.
3. Calcium Chloride solution was impregnated with sandy bed as absorber.
4. At night, the absorber is exposed to ambient air, when the desiccant collects moisture as shown in Fig. 2.7.

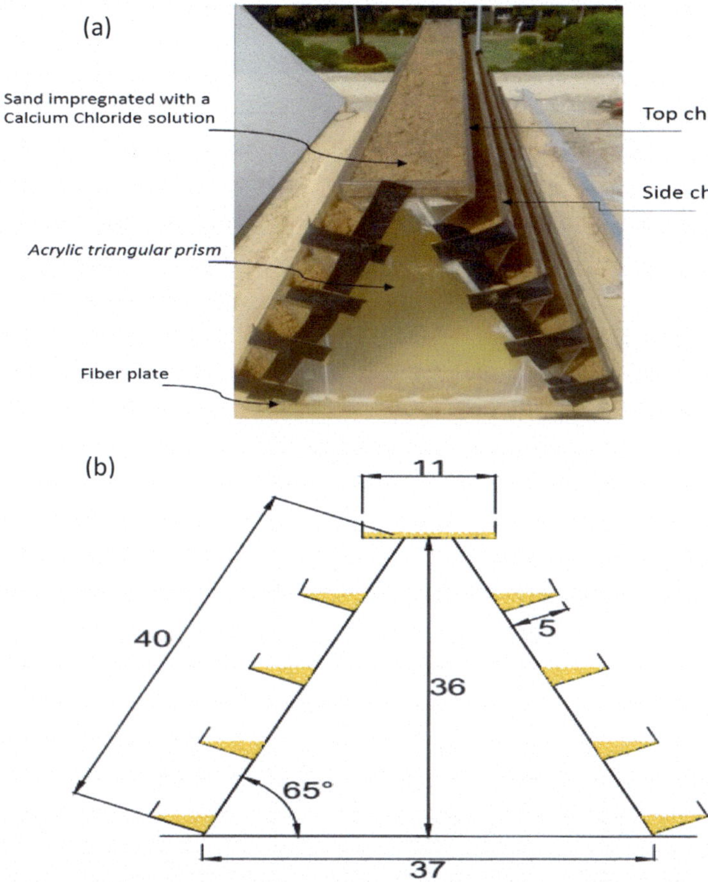

Fig. 2.5 **a** Photograph of the absorbing bed; and **b** schematic diagram of the bed (dimensions in cm)

5. At sunrise, the absorbing bed is covered with the condensation surface as shown in Fig. 2.8. The incident solar energy is absorbed by the absorbing bed, which gradually raises the temperature of the bed.
6. The vapor pressure difference between the solution at the bed and the condensation surface causes evaporation of vapor from the bed.
7. As shown in Fig. 2.9, vapor condenses at the condensation surface, which has a lower temperature.
8. The Condensate water was collected in graduated flask, which is connected with the sloped channels by a hose.
9. Water productivity, solar radiation intensity and temperatures of bed surface, condensation surface and air vapor are recorded through the day-time for east and west sides of the apparatus.

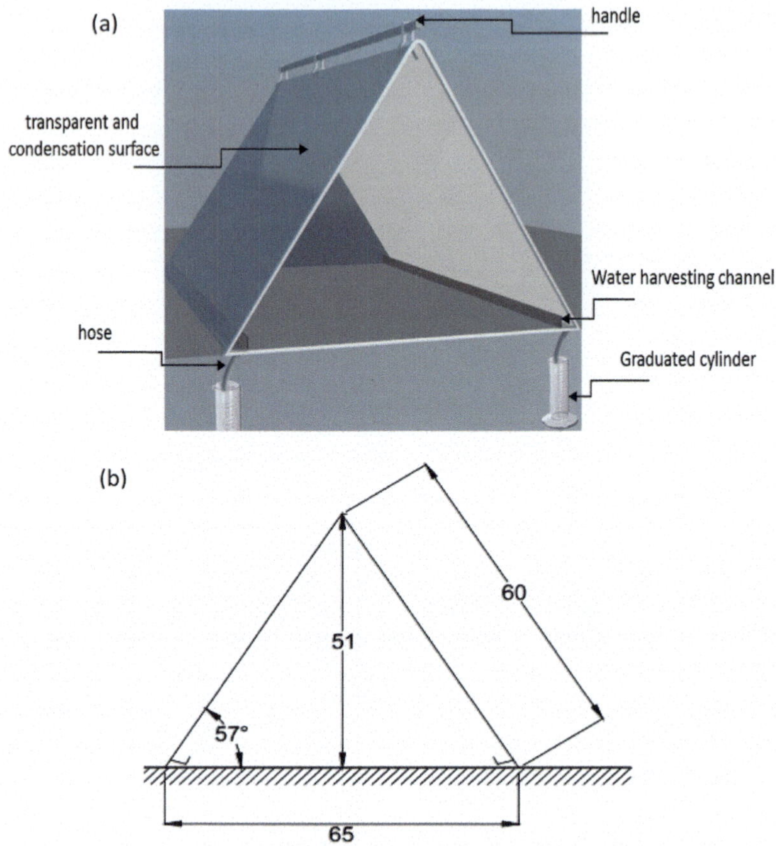

Fig. 2.6 a Perspective of transparent and condensation surface; and b schematic of transparent and condensation surface (dimensions in cm)

Fig. 2.7 Photograph of experimental apparatus at night

Fig. 2.8 Photograph of experimental apparatus at sunrise

Fig. 2.9 Photograph of experimental apparatus during the day time

10. The east side faces the incident solar radiation from sunrise to noontime and then, the west side faces it to sunset.
11. When the vapor pressure of the bed solution equals that of the condensation surface, evaporation and condensation terminate.
12. The experiment has been repeated every day in the periods (01–02/08), (17–21/08), (18–22/09) and (13–17/10).

2.6 Mathematical Model

2.6.1 Productivity Model

The system consists of bed and cover. During the period from sunrise to afternoon, the incident solar radiation with an intensity (H) passes through the apparatus as shown by dashed lines in Fig. 2.10. The following assumptions are made to develop the model:

1. One-dimension heat transfers under quasi-steady state condition.
2. Thermal properties of the system components are assumed constants.
3. During daytime operation, the system is completely locked off.
4. Condensation and evaporation processes occurring simultaneously.

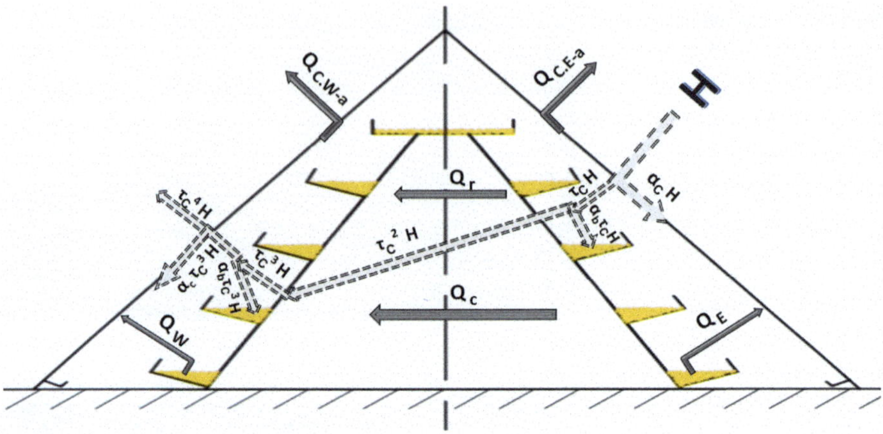

Fig. 2.10 Energy flow diagram from sunrise to noontime

5. Evaporated water condenses on the system surfaces having lower vapor pressure.
6. The air gap between bed and cover and the eastern and western bed is fixed. (Heat transfer is negligible).

The heat balance equation of east acrylic cover in the period from sunrise to afternoon can be expressed by the following relations corresponding to Fig. 2.10.

$$C_C \frac{dT_{C-E}}{dt} = (\alpha_C * H * A_{C-E} + Q_E) - Q_{CE-a} \tag{2.1}$$

where C_C refers to the cover's thermal capacity, which is calculated using the following formula;

$$C_C = m_C * c_{acry} \tag{2.2}$$

The sum of heat transfer through conduction, radiation, and evaporation from the east bed to the east cover is known as Q_E;

$$Q_E = Q_{r-E} + Q_{c-E} + Q_{ev-E} \tag{2.3}$$

The heat transfers from the east bed to the east cover by conduction (Q_{c-E}) can be calculated by the following relation;

$$Q_{c-E} = K_a * n * A_b * \frac{T_{s-E} - T_{C-E}}{\Delta Y} \tag{2.4}$$

where K_a, n, A_b, ΔY and T_{s-E}, are thermal conductivity of air, number of beds at every side, surface area of bed, average thickness of air gab between cover and bed, and bed surface temperature at east side, respectively. The east side bed surface temperature can be expressed in terms of the average east bed temperature as follows:

$$T_{s-E} = T_{b-E} + \alpha_b * \tau_C * H\left(\frac{\delta_b}{2K_b}\right) \tag{2.5}$$

where δ_b and K_b, are the thickness of the bed and the thermal conductivity of the bed, respectively. Dunkel's model is suitable for evaluating heat transport from east bed to east cover by radiation (Gad et al. 2001).

$$Q_{r-E} = F_{b-C} * \sigma * n * A_b * \left[(T_{s-E} + 273)^4 - (T_{C-E} + 273)^4\right] \tag{2.6}$$

where $F_{b-c,}$ denotes the shape factor between the bed and the cover, and it is assumed to equal unity in this analysis;

$$Q_{ev-E} = 0.0061\left[(T_{b-E} - T_{C-E}) + \frac{(P_{b-E} - P_{C-E})(T_{b-E} + 273)}{(0.265 - P_{b-E})}\right]^{\frac{1}{3}} * (P_{b-E} - P_{C-E}) * L * n * A_b \tag{2.7}$$

where, P_{b-E} and P_{C-E} are the water vapor partial pressures at the east bed surface and cover, respectively. The water vapor's partial pressure at the cover can be considered to be water vapor's saturation pressure of corresponding to the east cover temperature which can be computed by the following equation (Kabeel 2006);

$$\log p = \left[-3.21254 + 3.13619 \times 10^{-2} * T - 1.22512 \times 10^{-4} * T^2 + 3.63841 \times 10^{-7} * T^3 - 5.67607 \times 10^{-10} * T^4\right] \tag{2.8}$$

The partial water vapor pressure on the east bed (p_{b-E}) can be computed depending on solution concentration and average temperature of the bed in (mmHg) within a temperature range from 10 °C to 65 °C and concentration from 20 to 50% according to the following relations (Hamed 2000);

$$\ln p = a(X) - \frac{b(X)}{T + 111.96} \tag{2.9}$$

where a(X) and b(X) are regression parameters in terms of solution concentration;

$$a(X) = 10.0624 + 4.4674X \tag{2.10}$$

$$b(X) = 739.828 + 1450.96X \tag{2.11}$$

The following equation can be used to compute the water vapor partial pressure at the bed for a temperature range from 60 to 100 °C.

$$\ln p = a - \frac{b}{T + 273} \tag{2.12}$$

a and b, are defined in terms of concentration as given by Gad et al. (2001). The concentration of a solution is defined as the ratio of the mass of dry salt ($m_{d.s}$) to the gross mass of solution (m_{sol}), which equals the sum of the masses of water and dry salt:

$$X = \frac{m_{d.s}}{m_{sol}} \tag{2.13}$$

$$m_{sol} = m_{d.s} + m_w \tag{2.14}$$

The instantaneous value of a mass of solution can be computed from the following relation:

$$m_{sol(i)} = m_{sol(i-1)} - m_{evap(i-1)} \tag{2.15}$$

where $\boldsymbol{m_{evap}}$ is the mass of vaporized water from solution and the subscripts ($\boldsymbol{i}$) and ($\boldsymbol{i} - 1$) refer to the instantaneous and previous values, respectively.

$$m_{evap} = \frac{Q_{ev-E}}{L} \tag{2.16}$$

The initial value of the solution at the beginning of the experiment can be evaluated as a function of the solution's initial concentration:

$$m_{sol(0)} = \frac{m_{d.s}}{X_0} \tag{2.17}$$

where X_0 is an initial concentration of solution which can be calculated as expressed in Hamed (2000) by;

$$X_0 = \frac{\ln p_v - \left(a_o - \frac{b_o}{T_b + 119.6}\right)}{\left(a_1 - \frac{b_1}{T_b + 119.6}\right)} \tag{2.18}$$

where: $a_o = 10.0624$, $b_o = 739.828$, $a_1 = 4.4674$, $b_1 = 1450.96$

p_v is a water vapor pressure which can be assumed to equal a water vapor saturation pressure relative to ambient air temperature (T_a) at start of experiment as given in Eq. (2.8).

$$Q_{CE-a} = h_{C-a} * A_c(t_{C-E} - t_a) + F_{C-sky}\sigma A_C\left[(t_{C-E} + 273)^4 - (t_a + 273)^4\right] \tag{2.19}$$

where, F_{c-sky} is the factor of shape between the cover and surroundings and it is considered equal unity. A_C, is the surface area of east cover, h_{C-a} is the coefficient of convection heat transfer between the cover and the surrounding air and it can be expressed as a function of wind speed as evaluated by Mac Adam's relation as follows (Moffat 1988);

$$h_{C-a} = a + bv^n \tag{2.20}$$

where, a, b and n are constants dependent on the cover roughness and speed of the wind, and v is the speed of wind in (m/s) (Elsayed 1983). Referring to Fig. 2.10, the east bed heat balance can be expressed in the period from sunrise to afternoon as follows;

$$C_{b-E}\frac{dT_{b-E}}{dt} = (\alpha_b * \tau_C * H * n * A_b) - (Q_E + Q_r + Q_c) \tag{2.21}$$

where, $C_{b-E}$is the thermal capacity of the system in the east bed, which consists of sand, $CaCl_2$ salt, acrylic bed and water. It can be calculated by the sum of mass and specific heat product of system elements as given in the following relation;

$$C_{b-E} = m_C * c_{acry} + m_{w-E} * c_w + m_{sand-E} * c_{sand} + m_{CaCl_2-E} * c_{CaCl_2} \tag{2.22}$$

Heat transfer by conduction (Q_c) and by radiation (Q_r) from the east bed to the west bed can be expressed by;

$$Q_c = K_a * A_{s.b} * \frac{T_{b-E} - T_{b-W}}{d} \tag{2.23}$$

$$Q_r = \sigma * A_{s.b} * \left[(T_{b-E} + 273)^4 - \left(T_{b-W} + 273\right)^4\right] \tag{2.24}$$

Referring to Fig. 2.10, the west cover heat balance can be expressed in the period from sunrise to afternoon as follows:

$$C_C\frac{dT_{C-W}}{dt} = \left(\alpha_C * \tau_C^3 * H * A_C + Q_W\right) - Q_{CW-a} \tag{2.25}$$

where, Q_W, is the total amount of heat transferred from the west bed to the west cover by conduction, radiation, and evaporation.

$$Q_W = Q_{r-W} + Q_{c-W} + Q_{ev-W} \tag{2.26}$$

(Q_{r-W}), (Q_{c-W}), (Q_{ev-W}) and (Q_{CW-a}), can be calculated by modifying the parameters of a previous equations from (2.3) to (2.19) from east side to west side. Referring to Fig. 2.10, the west bed heat balance can be expressed in the period from the sunrise to the afternoon as follows:

$$C_{b-W}\frac{dT_{b-W}}{dt} = \left(\alpha_b * \tau_c^3 * H * n * A_b + Q_r + Q_c\right) - (Q_W) \quad (2.27)$$

During the period from afternoon to sunset, the incident solar radiation with an intensity (H) path change to direct from west to east side. The following relations can be used to express the heat equation of the cover and bed on the west and east sides during this period;

$$C_C\frac{dT_{C-E}}{dt} = \left(\alpha_C * \tau_C^3 * H * A_C + Q_E\right) - Q_{cE-a} \quad (2.28)$$

$$C_C\frac{dT_{C-W}}{dt} = (\alpha_C * H * A_C + Q_W) - Q_{cW-a} \quad (2.29)$$

$$C_{b.E}\frac{dT_{b-E}}{dt} = \left(\alpha_b * \tau_C^3 * H * n * A_b\right) - \left(Q_E + Q_r + Q_c\right) \quad (2.30)$$

$$C_{b.W}\frac{dT_{b-W}}{dt} = (\alpha_b * \tau_C * H * n * A_b + Q_r + Q_c) - (Q_W) \quad (2.31)$$

Amount of water evaporated from every bed can be calculated by equations;

$$\dot{m}_{evap-E(i)} = \frac{Q_{ev-E(i)}}{L} \quad (2.32)$$

$$\dot{m}_{evap-W(i)} = \frac{Q_{ev-W(i)}}{L} \quad (2.33)$$

If $T_{b-E} > T_{b-W}$

$$V_T = \sum_{i=1}^{z} \frac{Q_{ev-E(i)}}{L} * \left(\frac{A_C}{A_{cond}}\right) \quad (2.34)$$

If $T_{b-E} < T_{b-W}$

$$V_T = \sum_{i=1}^{z} \frac{Q_{ev-W(i)}}{L} * \left(\frac{A_C}{A_{cond}}\right) \quad (2.35)$$

where, A_C is the cover surface area and A_{cond} is the total condensation surface area including the area of the cold side bed. This ratio is used as a correction factor of productivity because it is found that some of the evaporated water condenses on the bed surface with lower temperature.

2.6.2 Model of Solar Radiation

The overall solar radiation incident on horizontal surface can be computed by the following relation:

$$H = H_B + H_d \tag{2.36}$$

where H_B is a horizontal surface beam radiation, which can be evaluated as a function of solar altitude angle (α) and normal incidence beam radiation (H_{Bn}) as follows:

$$H_B = H_{Bn}\sin(\alpha) \tag{2.37}$$

$$H_{Bn} = A\exp\left(\frac{-B}{\sin(\alpha)}\right) \tag{2.38}$$

where A and B, are solar radiation that is apparent at air mass zero (W/m^2) and coefficient of extinction in the atmosphere (dimensionless), respectively. Solar altitude angle (α) can be expressed as function of latitude (L_A), declination (δ) and hour (h) angles as shown below;

$$\sin\alpha = (\sin\delta * \sin L_A) + (\cos h * \cos L_A * \cos\delta) \tag{2.39}$$

$$\delta = 23.4\left[\sin\left(\frac{360 * (284 + N)}{365}\right)\right] \tag{2.40}$$

$$\mathrm{h} = \pm\frac{1}{4}(\text{Number of minutesfrom local solar noon}) \tag{2.41}$$

where it is assumed to be a positive value from noon to sunset and negative from sunrise to noontime. Latitude angle is taken for our location (Mansoura, Egypt) about (31 °N). The diffuse sky radiation (H_d) in Eq. (2.36) can be calculated by;

$$H_d = C_d * H_{Bn} * f_{ss} \tag{2.42}$$

where C_d and f_{ss}, are a diffuse radiation factor and factor of angle between the sky and the surface, respectively.

The system efficiency can be described as the ratio of heat required to evaporate water (Q_{evap}) to cumulative incident radiation (H) as follows:

The system efficiency can be described as a function of heat required to evaporate water (Q_{evap}) and the accumulated radiation incident (H) as follows:

$$\zeta = \frac{\sum Q_{evap}}{H_T} \tag{2.43}$$

Table 2.1 System design parameters

Symbol	Design parameter	Value
A_b	Surface area of bed per one channel	0.1 m^2
A_C	Cover area at west/east side	2.1 * 0.65 m^2
$A_{s.b}$	Area of bed side surface	0.01 m^2
a, b	empirical constants of equations	5.61, 1.09
c_{acry}	Specific heat of acrylic cover	1470 $\frac{J}{kg\ ^\circ C}$
c_w	water specific heat	4180 $\frac{J}{kg\ ^\circ C}$
c_{sand}	sand Specific heat	800 $\frac{J}{kg\ ^\circ C}$
c_{CaCl_2}	Calcium chloride specific heat	3060 $\frac{J}{kg\ ^\circ C}$
d	Average thickness of air gap between east and west bed	20 cm
K_a	Thermal conductivity of air	0.027 W/(m K)
n	Number of beds, Empirical constant	5, 1
α_C	Absorptivity coefficient of acrylic cover	0.04
α_b	Absorptivity coefficient of bed	0.87
ΔY	Average thickness of air gab between cover and bed	14 cm
ρ_C	Density of acrylic cover	1190 kg/m^3
τ_C	Transmissivity coefficient of acrylic cover	0.92
$m_{sand-E} m_{sand-W}$	Mass of dry salt at west/east side	4 kg/side
$m_{CaCl_2-W}, m_{CaCl_2-E}$	Mass of dry salt at west/east side	1.48 g/side

The previous equations are solved by numerically. Euler method is applied for solving the equations by taking one minute as a step through a day-time. The computer program evaluates the system transient parameters; temperature of cover and bed, water partial pressure at bed and cover, heat transfer by conduction, evaporation and radiation from bed to cover and concentration of solution at the east and west sides every minute. Then, performance parameters; productivity, water evaporation and system efficiency are evaluated. Parameters of system design are given in Table 2.1.

2.6.3 *Simulation*

The algorithm is carried out as follows:

1. Calculate solar radiation intensity (H) using solar radiation model Eqs. (2.35)–(2.41) corresponding to the day number.
2. Recording ambient temperature, and wind speed from metrological site.
3. Assume initial temperature of cover and bed at east and west sides equal ambient temperature.

4. Assume a bed in equilibrium with ambient air at initial conditions, and a water vapor pressure is assumed to equal a water vapor saturation pressure related to the temperature of ambient air (T_a) as given in Eq. (2.7).
5. Calculate initial concentration of abed solution from Eq. (2.18).
6. Compute the water vapor partial pressure at the bed surface as saturated pressure from Eq. (2.7) and cover from Eqs. (2.7)–(2.14).
7. Calculate Qc−E, Qr−E, Qev−E using Eqs. (2.3)–(2.6) and QCE−a using Eqs. (2.18) and (2.19).
8. Calculate CC and Cb−E using Eqs. (2.1) and (2.21), respectively.
9. Calculate Qc and Qr using Eqs. (2.22) and (2.23), respectively.
10. Calculate productivity using Eqs. (2.23)–(2.34) and concentration using Eqs. (2.12)–(2.14).
11. For the period from sunrise to afternoon, calculate temperature of east cover and bed using Eqs. (2.1) and (2.20), respectively.
12. Repeat previous steps for west side.
13. For the period from afternoon to sunset, calculate temperature of east cover and bed and west cover and bed using Eqs. (2.27)–(2.30).

2.7 Results and Discussion of Experimental and Theoretical Model

Experiments have been carried out in different months (August, September and October) at different ambient conditions, which results in clear changes in the accumulative productivity. Figure 2.11a displays the accumulative water productivity per day for all experimental days. It is clear that the month of August has the highest water productivity, while October has the lowest water productivity, corresponding to accumulative solar radiation intensity as shown in Fig. 2.11b.

Figure 2.12 shows the temperatures of condensation surface (T_C), air-vapor mixture (T_V) and bed (T_b) on the west and east sides, as well as the solar radiation for the period (15–17/10). It can be noticed that the solar radiation gradually rises from sunrise to its optimum value at afternoon, then decreased gradually as well as all temperatures gradually increased from sunrise to afternoon, then gradually decreased until sunset. From sunrise to afternoon, the temperature at the east side is higher than the west side temperature where the east side faces the sun. Then, the west side faces the sun beams and its temperature becomes higher than the east side.

Figure 2.13 shows a comparison between numerical and experimental results for system temperatures. It's observed that the theoretical bed temperature close to experimental bed temperature throughout the day-time except the period around noon time. Theoretical and experimental cover temperatures are close at every side except in the period from afternoon to the time before sunset when the west cover faces sun. In the period from sunrise to the time before afternoon, when the east cover faces sun, its cover temperature has some divergence in the values between experimental

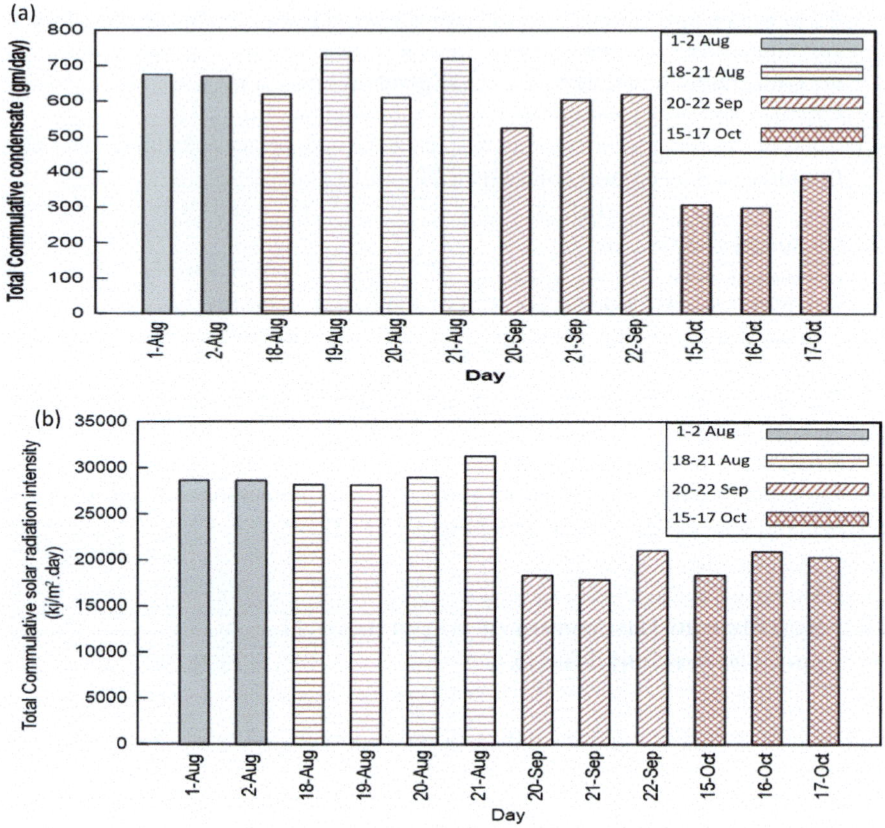

Fig. 2.11 **a** The accumulative water productivity per day for all test days; and **b** the accumulative solar radiation per day for all test days

and numerical. The theoretical and experimental solar radiation intensity values are shown in Fig. 2.14.

Figure 2.15 shows the cumulative productivity with time. It can be noticed that numerical values of evaporated water are higher than the experimental ones. This is due to the losses of the evaporated water which condenses on the lower temperature surfaces in the system.

It is observed from Fig. 2.16, that evaporation heat transfer at east side is higher than west side in the period from sunrise to afternoon when the productivity at west side is more than east side and vice versa in the period from afternoon to sunset. Also, the temperature of east bed and cover are higher than west side at the same period of high quantity of evaporation heat transfer as shown in Fig. 2.17.

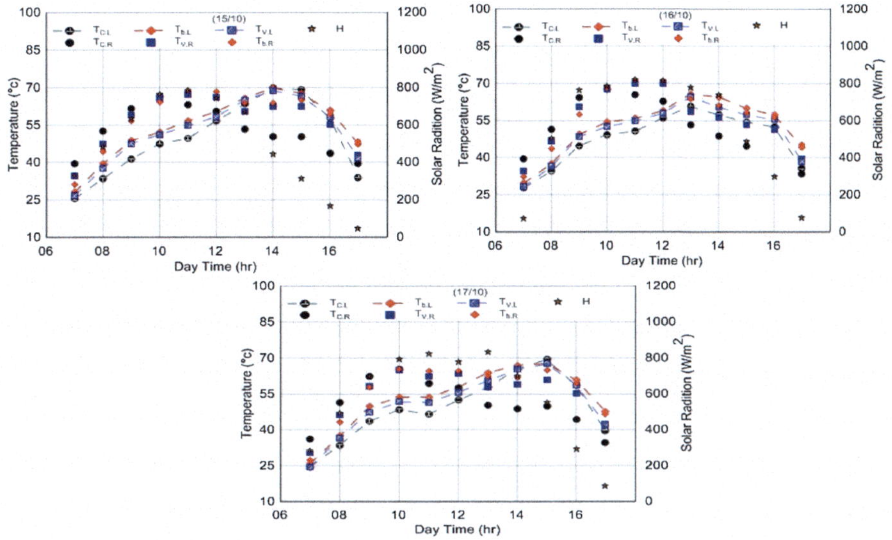

Fig. 2.12 Experimental hourly temperature and solar radiation

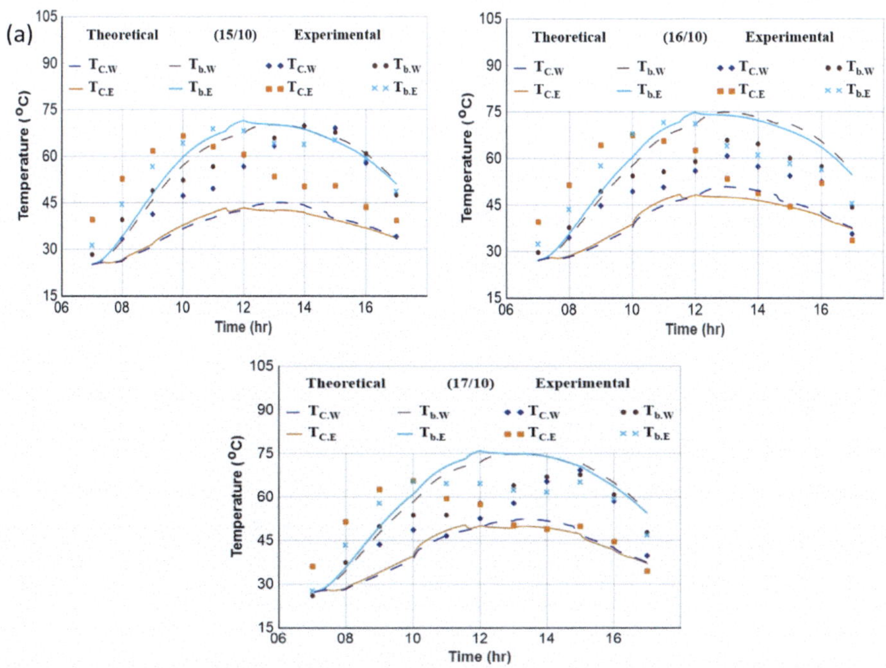

Fig. 2.13 Comparison between theoretical and experimental temperature

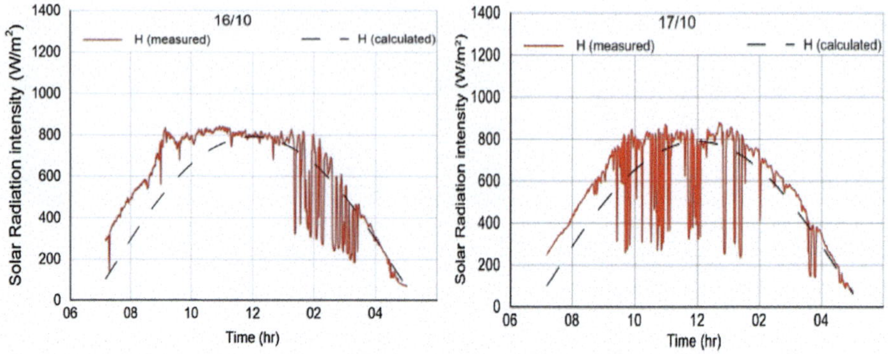

Fig. 2.14 Comparison between theoretical and measured solar radiation

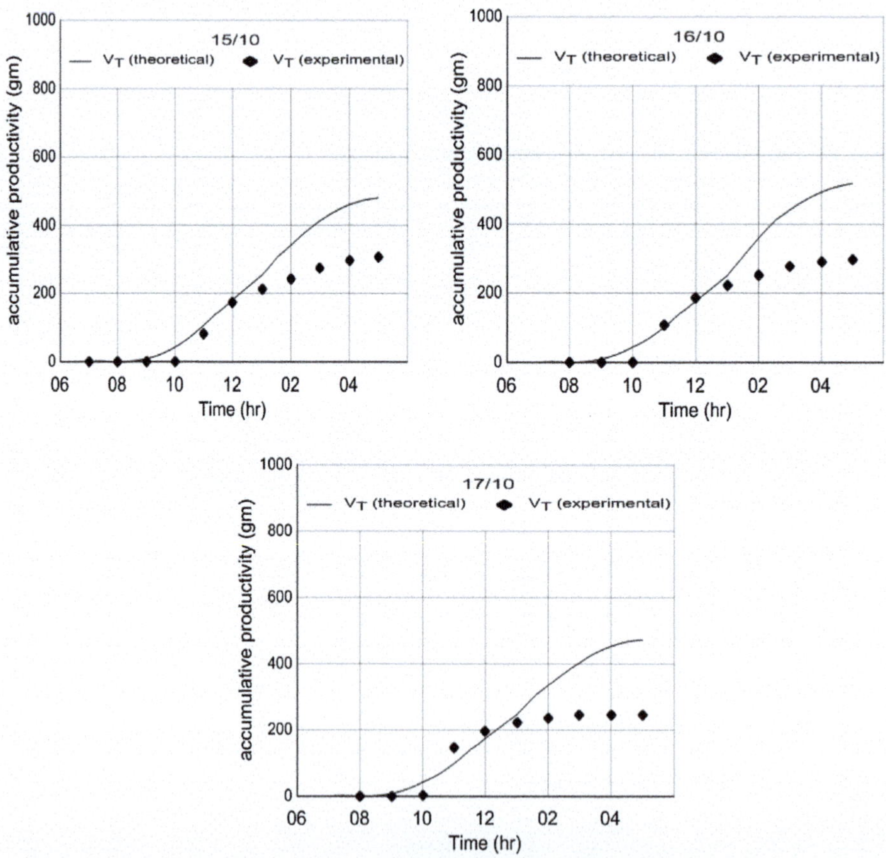

Fig. 2.15 Comparison between theoretical and experimental productivity and evaporated water

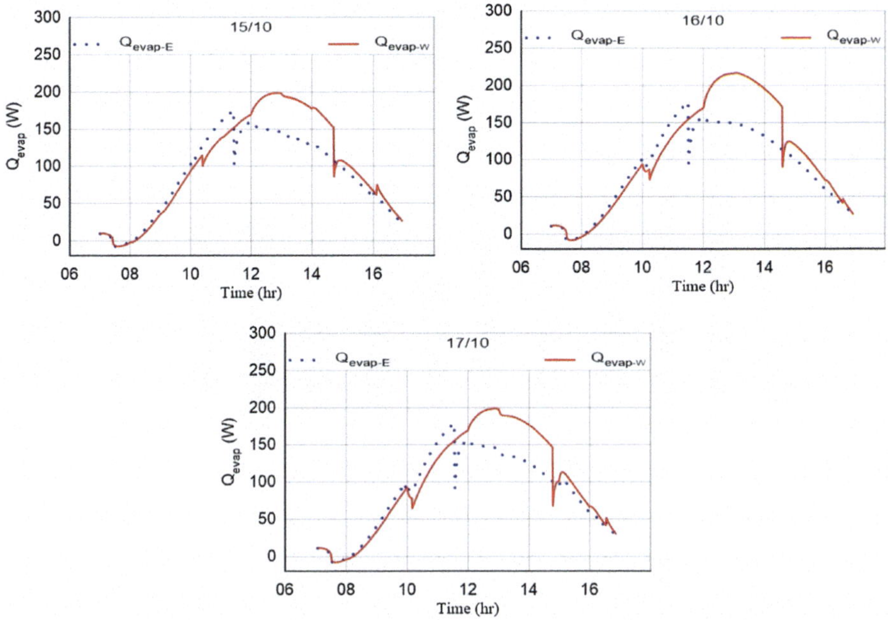

Fig. 2.16 Comparison between heat transfer by evaporation at east and west sides

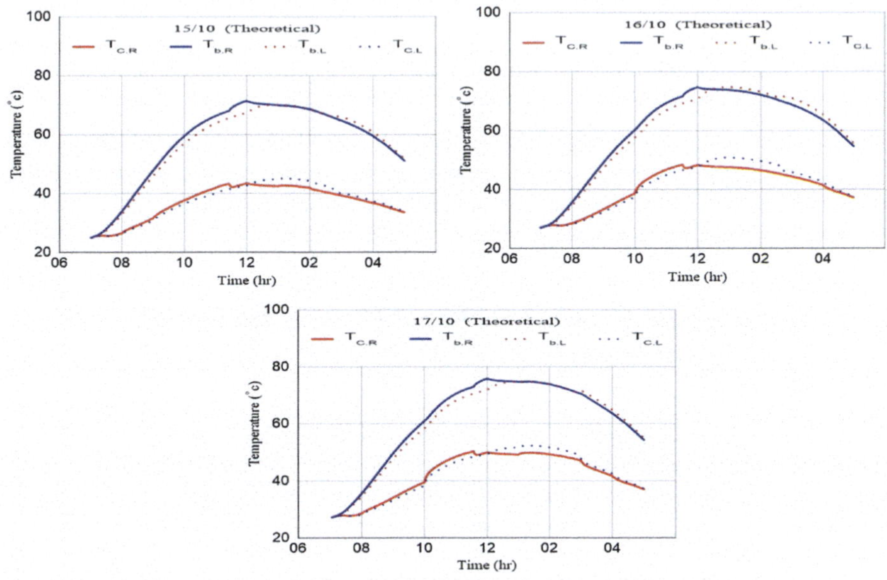

Fig. 2.17 Numerical values of bed and cover temperatures at east and west sides

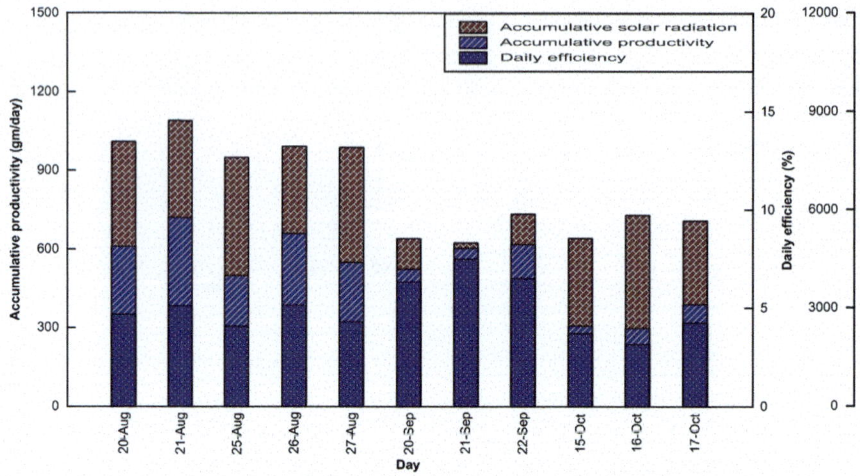

Fig. 2.18 Daily efficiency corresponding to accumulative solar radiation and productivity

2.7.1 *Daily Efficiency*

$$\eta = \frac{\text{heat of generated water}}{\text{cumulative solar radiation}}$$

Figure 2.18, shows the cumulative productivity, accumulative solar radiation and daily efficiency with time. The daily efficiency of the experiment lies between 3 and 7%. It doesn't depend only on accumulative productivity, but it is also inversely proportional to the corresponding accumulative solar radiation. It can be noticed that the system's daily efficiency is low because of the lowering value of productivity compared to the corresponding accumulative solar radiation intensity.

2.7.2 *Comparison Between Different Designs Including the Present Work*

Table 2.2 shows a comparison of bed type and productivity for previous study results conducted on extracting water from moist air using desiccant. It's obtained that our accumulated productivity less than some of these studies and this due to some reasons as follows:

1. Using sandy bed without any improvement.
2. Some of evaporated water does not condense on the surface of the collector due to the difference of temperature between east and west bed.
3. Trapped air in the gap between the bed and cover.

Table 2.2 A review of previous research on desiccant water extraction from moist air

Researcher(s)	Type of desiccant bed	Amount of water	Experiment place and date
Alayli et al. (1987)	S-shaped isotherms of adsorbent composite material	1.0 l/day m^2	Algeria (1987)
Gad et al. (2001)	Corrugated bed was made of cloth ($CaCl_2$ as desiccant)	1.5 l/day m^2	Mansoura—Egypt (2001)
Kabeel (2006)	Sandy bed impregnated with $CaCl_2$ at about 30% concentration	1.2 l/day m^2	Tanta—Egypt (2006)
Kabeel (2007)	Capability of two glass pyramid shapes with a multi-shelf solar system of the same dimensions The first pyramid's bed was constructed of saw wood The bed of the second pyramid was composed of cloth In comparison to the cloth bed, the saw wood bed absorbs less solution	2.5 l/day m^2	Tanta—Egypt (2007)
Ji et al. (2007)	A composite adsorbent was used (host matrices of ultra-large pore crystalline material MCM-41 and hygroscopic salt calcium chloride)	More than 1.2 l/day m^2	China (2006)
Hamed et al. (2011)	A calcium chloride ($CaCl_2$) impregnated sandy bed	1.0 l/day m^2	Taif area, Saudi Arabia (2011)
Kumar and Yadav (2015)	"$CaCl_2$/saw wood" as a desiccant	0.5 l/day m^2	NIT Kurukshetra—India (2015)
William et al. (2015)	Using $CaCl_2$ at initial saturation concentration (30%) as desiccant trapezoidal prism solar collector compare between cloth and sand as host materials	310 (for sand bed at 19-September) 0.870 (for cloth bed at September)	Egypt (2015)

(continued)

Table 2.2 (continued)

Researcher(s)	Type of desiccant bed	Amount of water	Experiment place and date
Wang et al. (2017)	The corrugated filling mode of consolidated active carbon felt mixed with LiCl sorbent is invented	Harvesting 12.5 to 14.7 kg of water while packaging 40.8 kg of sorbents	Shanghai, China (2016)
Talaat et al. (2018)	They developed a portable device using calcium chloride ($CaCl_2$) solution as desiccant	An average of 0.33–0.63 kg/m^2/day is harvested	Mansoura, Egypt (2018)
Srivastava and Yadav (2018)	[LiCl-sand (CM-1)], [$CaCl_2$-sand (CM-2)] and [LiBr/sand (CM-3)] were applied as salts with a 37% concentration and sand as a host material designed 1.54 m^2 Scheffler reflector	CM-1 … 0.90 ml/day CM-2 … 0.115 ml/day CM-3 … 0.73 ml/day	Kurukshetra, Haryana, India (2018)
Elashmawy and Alshammari (2020)	They used a parabolic sun concentrator to activate the tubular solar still, which increases its ability to evaporate water from a strong desiccant of calcium chloride at extremely low humidity conditions	Harvesting 0.5 L of water per kg of calcium chloride	Hail city (27.64 °N, 41.75 °E), Saudi Arabia (2019)
Fathy et al. (2020)	They developed a foldable device using calcium chloride ($CaCl_2$) solution as desiccant	272–750 g/day	Mansoura, Egypt (2020)
Sleiti et al. (2021)	A device has been introduced that includes silica gel as an absorbent material subjected to heat flux, unit of a water sorbent, a reflector, and a condenser	159 g/kg of silica gel	Qatar (2021) controlled indoor environment under the conditions of 22 °C and (RH) from 30 to 60%

(continued)

Table 2.2 (continued)

Researcher(s)	Type of desiccant bed	Amount of water	Experiment place and date
Li and Zhang (2022)	Water harvesting apparatus was introduced, which consisted of: activated carbon thin layer film of polyethylene (commercially available) container for water	1.13 kg/m^2/day	China (2022) Very dry regions (air RH < 10%, soil moisture content < 3%) Interfacial solar heating under natural sunshine

2.8 Conclusion

Extracting water from air by desiccant using oblique condensation surface and solar energy as a heating source is presented. From the experimental study and numerical investigation, the following conclusions could be drawn:

1. The water productivity is dependent on ambient conditions and a maximum value of about 735 gm/m^2 day is recorded.
2. The difference between the temperature of condensation surface and air-vapor has a great effect on water productivity.
3. The system has low daily efficiency where its maximum value is about 7%.
4. Orientation of bed surface must be selected such that the bed temperature has the same values for most of the operation period.

References

Alayli Y, Hadji NE, Leblond J (1987) A new process for the extraction of water from air. Desalination 67(C):227–229. https://doi.org/10.1016/0011-9164(87)90246-3

Elashmawy M, Alshammari F (2020) Atmospheric water harvesting from low humid regions using tubular solar still powered by a parabolic concentrator system. J Clean Prod 256:120329

Elsayed MM (1983) Comparison of transient performance predictions of a solar-operated diffusion-type still with a roof-type still

Fathy MH, Awad MM, Zeidan EB, Hamed AM (2020) Solar powered foldable apparatus for extracting water from atmospheric air. Renew Energy 162. https://doi.org/10.1016/j.renene.2020.07.020.

Gad HE, Hamed AM, El-Sharkawy II (2001) Application of a solar desiccant/collector system for water recovery from atmospheric air. Renew Energy 22(4):541–556. https://doi.org/10.1016/S0960-1481(00)00112-9

Hamed AM (2000) Absorption-regeneration cycle for production of water from air-theoretical approach. Renew Energy 19(4):625–635. https://doi.org/10.1016/S0960-1481(99)00068-3

Hamed AM, Aly AA, Zeidan E-SB (2011) Application of solar energy for recovery of water from atmospheric air in climatic zones of Saudi Arabia. Nat Resour 2(01):8

Ji JG, Wang RZ, Li LX (2007) New composite adsorbent for solar-driven fresh water production from the atmosphere. Desalination 212(1–3):176–182. https://doi.org/10.1016/j.desal.2006.10.008
Kabeel AE (2006) Application of sandy bed solar collector system for water extraction from air. Int J Energy Res 30(6):381–394. https://doi.org/10.1002/er.1155
Kabeel AE (2007) Water production from air using multi-shelves solar glass pyramid system. Renew Energy 32(1):157–172. https://doi.org/10.1016/j.renene.2006.01.015
Kumar M, Yadav A (2015) Experimental investigation of solar powered water production from atmospheric air by using composite desiccant material 'CaCl2/saw wood.' Desalination 367:216–222. https://doi.org/10.1016/j.desal.2015.04.009
Li L, Zhang J (2022) Water harvesting from desert soil via interfacial solar heating under natural sunlight. J Colloid Interface Sci 607:1986–1992. https://doi.org/10.1016/J.JCIS.2021.09.195
"Market research reports, consulting: Global Market Insights Inc. (2022) https://www.gminsights.com/. Accessed 05 May 2022
Moffat RJ (1988) Describing the uncertainties in experimental results. Exp Therm Fluid Sci 1(1):3–17. https://doi.org/10.1016/0894-1777(88)90043-X
"NASA Earth Observatory—Home." https://earthobservatory.nasa.gov/. Accessed 05 May 2022
Obrezkova VE (1988) Hydro-energy. Energoatomezdat, Moscow
Our Story|Maithri Aquatech (2022) https://www.maithriaqua.com/our-story. Accessed 06 Jun 2022
Product (2022) http://www.skywell.com/product/. Accessed 29 May 2022
Sleiti AK, Al-Khawaja H, Al-Khawaja H, Al-Ali M (2021) Harvesting water from air using adsorption material–Prototype and experimental results. Sep Purif Technol 257:117921
Srivastava S, Yadav A (2018) Water generation from atmospheric air by using composite desiccant material through fixed focus concentrating solar thermal power. Sol Energy 169:302–315
Talaat MA, Awad MM, Zeidan EB, Hamed AM (2018) Solar-powered portable apparatus for extracting water from air using desiccant solution. Renew Energy 119. https://doi.org/10.1016/j.renene.2017.12.050
United Nations World Water Assessment Program (fourth ed.) (2012) UN world water development report: managing water under uncertainty and risk, vol 1
Wang JY, Wang RZ, Wang LW, Liu JY (2017) A high efficient semi-open system for fresh water production from atmosphere. Energy 138:542–551
WatAir - Converting air to drinking water (2022) https://www.watairuk.co.uk/. Accessed 05 May 2022
Watergen|Water from Air (2022) https://www.watergen.com/. Accessed 05 May 2022
WaterMaker India (2022) http://watermakerindia.com/projects.php. Accessed 06 Jun 2022
William GE, Mohamed MH, Fatouh M (2015) Desiccant system for water production from humid air using solar energy. Energy 90:1707–1720. https://doi.org/10.1016/j.energy.2015.06.125

Chapter 3
New Materials for Sorption-Based Atmospheric Water Harvesting: Opportunities and Challenges

L. G. Gordeeva and M. V. Solovyeva

3.1 Introduction

Nowadays, due to population growth, climate change, and environmental pollution, freshwater scarcity is becoming one of the global challenges. Although the Earth's total freshwater resources of 35×10^6 km^3 are abundant, most of them are hard-to-reach glaciers, permanent snow cover, and deep underground water (Shiklomanov 1993). The rest (rivers, fresh lakes, swamps) is unevenly distributed on the earth, making the shortage of freshwater a factor limiting the economic and social development of vast areas, particularly those located in arid regions. The areas of Central America, the west coastline of South America, North and South Africa, the Near and Middle East, India, North China, Mongolia, and Australia are classified as the regions with a high (water resources vulnerability index equal to the ratio of withdrawal to supply of 40–80%) and extremely high stress (the index > 80%) (Progress on level of water stress, Global Status and Acceleration Needs for SDG Indicator 2021).

Meanwhile, the atmosphere is a vast and accessible worldwide moisture source of 12.9×10^3 km^3 exceeding the total annual human demands for freshwater including domestic (food, sanitaria), agriculture, and industrial sectors (3000–4000 km^3) (Wada et al. 2016). Even in the aridest regions of the earth (the Sahara Desert, the Mojave Desert, Saudi Arabia and Central Australia) the moisture content of the atmosphere reaches 4–15 g/m^3 (Griffiths 1972). For this reason, water harvesting from the atmosphere has attracted mankind's attention since ancient times. Artificial dew ponds were constructed in England in the Middle Ages. Bowl-shaped hollows, covered by dry straw and clay dug in the ground were used to collect dew in the nighttime.

L. G. Gordeeva (✉) · M. V. Solovyeva
Boreskov Institute of Catalysis, Ac. Lavrentiev Av. 5, Novosibirsk, Russia
e-mail: gordeeva@catalysis.ru

M. V. Solovyeva
e-mail: solovyeva@catalysis.ru

E. Fosso-Kankeu et al. (eds.), *Atmospheric Water Harvesting Development and Challenges*, Water Science and Technology Library 122,
https://doi.org/10.1007/978-3-031-21746-3_3

Several legends tell about the so-called dew "spring" built in the Touran steppe in Altai Mountain (ex-USSR) (Beysens and Milimouk 2000). So-called "fountains" or big reservoirs of water, which presumably were built in the Middle Ages and served as dew collectors, were found by a forester and engineer F.I. Zibold near the city of Feodosia in the Crimean Peninsula in 1905 (Kogan and Trahtman 2003). Although most of these legends can hardly be proven due to the lack of reliable documents, they indicate that people have long tried to extract water from the atmosphere.

Nowadays, there are three ways of harvesting atmospheric moisture: fog and dew collection, and sorptive water extraction by solid or liquid desiccants (Klemm et al. 2012; Khalil et al. 2016; Tu et al. 2018). As shown above, the two first methods have quite a long history of development. Nowadays the feasibility of these methods of atmospheric water harvesting is successfully demonstrated by a large number of commercial atmospheric water generators and fog collection plants built in different areas around the world (Salehi et al. 2020). On the contrary, the sorptive method is less developed. To the best of our knowledge, the Adsorption Water Harvesting from the Atmosphere (AWHA) was suggested for the first time in 1934 by Altenkirch (1934). He proposed gaining water by exposing a hygroscopic substance to the air overnight to sorb moisture, then heating it with solar energy to desorb vapor and cooling it to condense water. Elmer and Hyde used hygroscopic salts LiCl, $CaCl_2$, $MgCl_2$, and $NaC_2H_3O_2$ supported on various carriers (microporous glass, fibrous board, and sand) to extract moisture from the air at night-time and desorb/collect water during the day using solar heat (Elmer and Hyde 1986). The amount of water adsorbed reached 0.24–0.62 L/m^2 at relative humidity (RH) of the air of 50%; at heating the desiccant to the temperature of 110–116 °C about 80% of adsorbed water was released. Alayli et al. (1987) studied the AWHA system in daily cycles with the adsorption during the nocturnal phase at RH = 20–80% and solar heat-driven desorption during the diurnal phase at a temperature of 100 °C, followed by vapor condensation on a cold plate. The amount of water collected using composite sorbents varied from 1 to 4 L/m^2 of the adsorbent at RH = 50–80%.

The properties of the desiccant, including equilibrium and dynamics of water adsorption, and particularly their matching to the climatic conditions of the specific region, the temperature grade of the driving heat source, and system components configuration are key factors affecting the performance of AWHA (Tu et al. 2018; LaPotin et al. 2019; Asim et al. 2021). Nowadays, a huge number of various sorbents have been developed; thus, the Handbook of porous solids published in 2002 (Schüth et al. 2002) contains five volumes. The properties of some adsorbent materials can be rationally designed to harmonize them with the requirements of the specific cycle. The achievements of material science triggered the resurgence of research interest in AWHA.

This chapter addresses the progress and challenges of AWHA with a particular focus on the desiccants offered for this application. First, the basic principles and performance indexes of AWHA are described. Then, we consider what sorbent is required for AWHA; the thermodynamic and dynamic requirements for properties of the needed adsorbent are discussed. The desiccants both traditional and innovative suggested for AWHA are described in terms of meeting their properties to the formulated requirements. Then, the main cycles and system configurations are briefly considered. Finally, the challenges and outlooks are discussed.

3.2 The Basic Principle of AWHA

It's known that a crucial issue for the traditional methods of atmospheric water harvesting such as fog, dew, and rain collection is their strong dependence on the geographic location and climatic conditions, particularly, RH of the air (Tu et al. 2018; Gido et al. 2016; Domen et al. 2014). Therefore, the sorption method was developed to alleviate this dependence and to increase the amount of water harvested by using the desiccant. Compared with the dew and fog collection, the sorbent-assisted approaches can be realized in a much wider RH range and exhibits larger water collection capacities.

The AWHA method utilizes the natural daily variations in RH that are typical of arid regions. In fact, the absolute humidity of air remains about constant during the day (Fig. 3.1a). Wide swing between day and night temperatures leads to a large variation in RH, which can reach 15–30% in desert regions in Africa, Central Australia, and Saudi Arabia (Fig. 3.1c). The simplest daily AWHA cycle consists of three main stages (Fig. 3.2a) (Alayli et al. 1987):

1. Sorption of the atmospheric moisture on the dried sorbent at night when the air RH is high. The sorption stage can be isenthalpic (line $A_{in} - A_{out}$ in Fig. 3.2b), if there is no heat exchange between the adsorber and the ambient, or isothermal if the adsorption heat is dissipated to the ambient.
2. Heating the desiccant bed and desorption of the stored water from the sorbent during the day-time. To desorb the captured vapor, heat has to be supplied to the sorbent; renewable solar energy available in abundance in arid regions is often considered the driving energy source (Ibrahim et al. 2018; Talaat et al. 2018). During the vapor desorption, the moisture content in the process air increases up

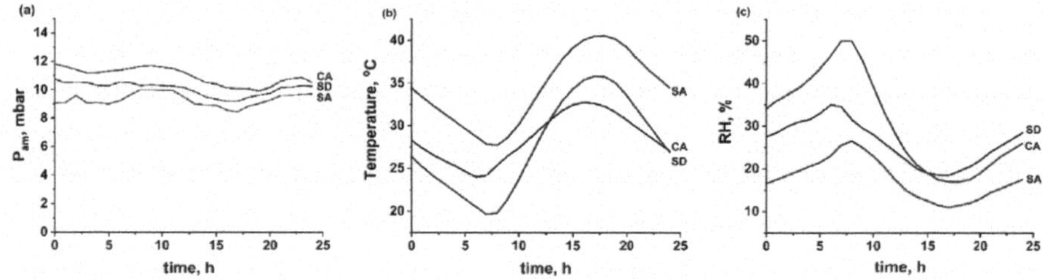

Fig. 3.1 Daily variation of the partial water vapor pressure in the air (**a**), temperature (**b**), and relative humidity RH (**c**) in the Sahara Desert (SD), Saudi Arabia (SA) (July, 10th), and Central Australia (CA) (January, 10th) [Reprinted from Gordeeva et al. (2020) Copyright 2019, with permission from Elsevier]

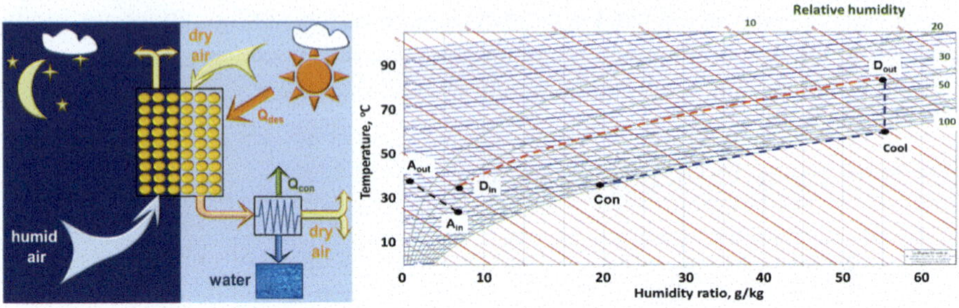

(a) (b)

Fig. 3.2 The scheme of AWHA (**a**) and AWHA cycle on the psychrometric chart of humid air (**b**): the process air state during adsorption ($A_{in} - A_{out}$), desorption ($D_{in} - D_{out}$), cooling and condensation (D_{out}-Cool-Con) stages

to the equilibrium value over the sorbent saturated by water during the preceding adsorption stage (line $D_{in} - D_{out}$);

3. Cooling of the process air down to the dew point (line D_{out} − Cool) and subsequent condensation of desorbed vapor (line Cool -Con) and collection of water in a tank.

The humidity content in the process air during the desorption stage increases up to a dew point higher than the ambient temperature (Fig. 3.2b), which allows the condensation of the desorbed vapor using the ambient as the heat sink for the condensation heat. Thus, contrary to dewing systems, the AWHA can be implemented even if the RH of the ambient air is very low without additional cooling of the air (Gordeeva et al. 2020).

The performance of the AWHA system can be characterized by the following indexes (Gordeeva et al. 2020; Wang et al. 2018):

- the specific moisture harvesting capacity of the sorbent

$$\Delta w[g/g] = w(T_{ad}, P_{am}) - w(T_d, P_{am}), \tag{3.1}$$

where $w(T_{ad}, P_{am})$ and $w(T_d, P_{am})$ are the equilibrium uptakes at temperatures T_{ad} of adsorption and T_d of desorption, respectively, and the ambient water vapor pressure P_{am};

- the specific water production per cycle SWP, which determines the mass of the sorbent needed to get, let's say, 1 L of water

$$SWP[g/g] = m_{con}/m_s, \tag{3.2}$$

where m_{con} and m_s denote the masses of the water condensed and the sorbent, respectively. SWP related per unit surface S_s of the sorbent exposed to the air might be used as well

$$SWP_S\left[\mathrm{g/m^2}\right] = m_{\mathrm{con}}/S_\mathrm{s}; \tag{3.3}$$

- the specific thermal energy consumption

$$\begin{aligned} SEC_{\mathrm{th}}\left[\mathrm{kJ/g}\right] &= Q_\mathrm{d}/m_{\mathrm{con}} \\ &= \left[\left(m_s C_{p,\mathrm{s}} + w_{ad} C_{p,\mathrm{w}}\right)(T_\mathrm{d} - T_{\mathrm{ad}}) + m_\mathrm{s} \Delta w \Delta H_\mathrm{d}\right]/m_{\mathrm{con}}, \end{aligned} \tag{3.4}$$

  where Q_d is the heat supplied for desorption, $C_{p,s}$ and $C_{p,\mathrm{w}}$ are the specific heats of the adsorbent and adsorbed water, $w_{\mathrm{ad}} = w(T_{\mathrm{ad}}, P_{\mathrm{am}})$, T_d and T_{ad} are the regeneration and adsorption temperatures ΔH_d is the desorption enthalpy;
- the efficiencies δ_{ex} of water extraction during adsorption stage, and δ_{col} of water collection during desorption/condensation

$$\delta_{\mathrm{ex}} = m_{\mathrm{ad}}/m_{\mathrm{w.in}} = (X_{\mathrm{Ain}} - X_{\mathrm{Aout}})/X_{\mathrm{Ain}}, \tag{3.5}$$

$$\delta_{\mathrm{con}} = m_{\mathrm{con}}/m_\mathrm{d} = (X_{\mathrm{Dout}} - X_{\mathrm{Con}})/X_{\mathrm{Dout}}, \tag{3.6}$$

  where m_{ad}, $m_{\mathrm{w.in}}$ are masses of water adsorbed and entering into the adsorber during adsorption, m_{con} and m_d are masses of water condensed and desorbed, X_{Ain}, X_{Aout}, X_{Dout}, and X_{Con} are the humidity ratio of air at points $\mathrm{A_{in}}$, $\mathrm{A_{out}}$, $\mathrm{D_{out}}$, and Con (Fig. 3.2b), respectively;
- recovery ratio or the ratio of the mass of water collected to the mass of water entering to the adsorber

$$\mathrm{RR} = m_{\mathrm{con}}/m_{\mathrm{w.in}} = \delta_{\mathrm{ex}}\delta_{\mathrm{con}}, \tag{3.7}$$

  assuming $m_{\mathrm{ad}} = m_\mathrm{d}$ at cyclic equilibrium process;
- thermal and primary energy efficiencies

$$\eta_{\mathrm{th}} = m_{\mathrm{con}} L/Q_\mathrm{d} \tag{3.8}$$

$$\eta_{\mathrm{PE}} = m_{\mathrm{con}} L/Q_{\mathrm{sol}} \tag{3.9}$$

  where L is the latent heat of water and Q_{sol} is solar energy.

For large-scale systems, the adsorbent has to be packed into a bed (Yang et al. 2021). The air blowing through the bed needs forced convection by the electric power. The electric energy consumption SEC_{el} for blowing air should be considered as well, which depends on the adsorbent bed configuration, humidity ratio of the air, and the recovery ratio RR. The importance of RR can be emphasized given the rather low moisture content of the air in arid regions. At a moisture content of 5–10 g/m^3 the volume of air to be processed can be calculated as 100–200 m^3/L water even at RR = 1. At decreasing RR, the air volume grows inverse proportionally.

To increase the productivity of the harvester, several adsorption/desorption cycles per day can be implemented (Hanikel et al. 2019). Then the specific daily production $SDWP = N \times SWP$ (N—number of sorption/desorption cycles per day) can be used. However, the *SWP* per cycle during daytime might decrease due to a smaller amount of water sorbed at a lower RH value. During night-time, an additional heat source for the desorption is required.

3.3 Desiccant Materials for AWHA

Desiccant materials can be categorized into aBsorbents and aDsorbents. According to McBain (2009), who coined the term "sorption", the aBsorption can be defined as the occlusion of the molecules through their diffusion inside the phase of the desiccant resulting in the formation of liquid or solid solution (Fig. 3.3 left). The aDsorption is the condensation of the adsorbate on the surface of the adsorbent (Fig. 3.3 center). Based on the interaction mechanisms, adsorption can involve physisorption and chemisorption. There are three basic contributions to the sorbent-sorbate interactions: dispersion, electrostatic and chemical bonds (Yang 2003). Physisorption comprises the first two forces, while chemisorption involves the formation of chemical bonds.

A large family of composite desiccants based on hygroscopic salts inserted inside porous adsorbents can be classified as an intermediate class of desiccants, as they trap moisture through adsorption, chemical reaction, and absorption (Fig. 3.3 right) (Gordeeva and Aristov 2012). Below these types of desiccant materials will be considered in light of their applicability to AHWA. But first, we will outline what adsorbent is needed for AWHA and discuss the properties of the adsorbent, which ensure high performance and sustainable operation of AWHA systems.

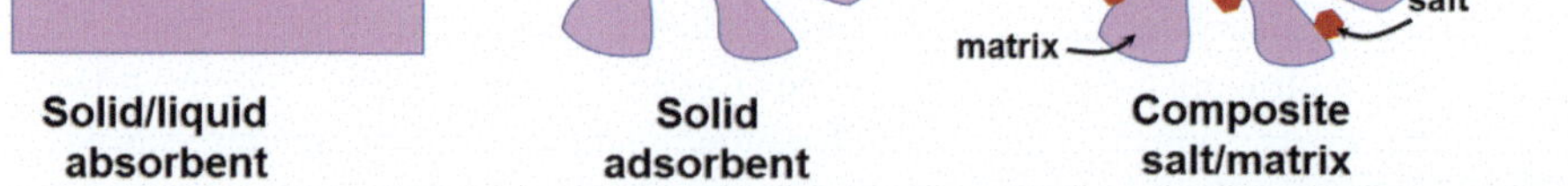

Fig. 3.3 Water sorption mechanisms: absorption by solids/liquids (left), adsorption on porous solids (centre), and adsorption, salt hydrate formation, deliquescence, and absorption by the salt solution (right)

3.3.1 Requirements for the Sorbent Properties

The sorbent is a key component of AWHA and its properties, and in particular, their agreement with the operating conditions of the AWHA cycle, the system components, and climatic conditions of the specific region, where AWHA is to be realized, is a prerequisite for the successful implementation of this technology. The water adsorption isotherm $w(P/P_0)|_{T=cons}$ is the basic characteristic of the adsorbent, which determines the performance of AWHA. There is no standardized criterion for the adsorbent selection for AWHA. Let's try to understand, what adsorbent is profitable for this application.

3.3.1.1 Thermodynamic Requirements

The basic thermodynamic requirements for the adsorbent properties specialized for AWHA can be derived based on the analysis of the effect of water adsorption isotherm on the performance of AWHA. It was shown *for a common flow adsorber* (Gordeeva et al. 2020; Solovyeva et al. 2021a), that, in order to effectively sorb moisture and dehumidify the ambient air during the adsorption stage, the sorbent must have a strong affinity to water vapor. In other words, it has to trap water at low RH to ensure a low humidity ratio X_{Aout} in the adsorber outlet and high fraction δ_{ex} of water extraction (Eq. 3.5). Such adsorbents are characterized by the isotherm of I type according to IUPAC (Thommes et al. 2015). On the contrary, for the desorption stage, the adsorbent with a low affinity, which easily releases the sorbed water at high RH is needed. Indeed, a high humidity ratio X_{Dout} can be reached over such an adsorbent, which is advantageous for increasing the collection efficiency δ_{col} (Eq. 3.6). Such adsorbents are characterized by stepped water adsorption isotherms of V type with uptake at high RH, corresponding to the adsorption stage. Thus, the choice of an adsorbent for AWHA is a matter of compromise: the adsorbent must possess both strong and weak adsorption sites, the strong sites ensure effective drying of the ambient air and a high fraction δ_{ex}, while weak ones promote a high efficiency δ_{col}.

The water uptake w depends on both, relative pressure and temperature. It was shown, that for many adsorbents, both micro- and mesoporous (Bering et al. 1966; Dubinin and Stoeckli 1980), composites salt/matrix (Gordeeva et al. 2007, 2008), and MOFs (Lange et al. 2015), there is a one-to-one correspondence between the uptake and the Polanyi-Dubinin adsorption potential $\Delta F = - RT \ln (P/P_0)$ (Polanyi 1932), $w = f(\Delta F)$. For this reason, ΔF can be used as the measure of the sorbent affinity to water vapor. To harmonize the adsorbent with the climatic conditions of a specific region and the temperature potential of the driving heat source, the quantitative requirements for the adsorbent properties for AWHA can be expressed as the values $\Delta F_{ad} = RT_{ad} \ln (P_0(T_{ad})/P_{am})$ and $\Delta F_d = RT_d \ln (P_0(T_d)/P_{am})$, corresponding to the conditions (water vapor pressure P_{am} and the temperatures T_{ad} and T_d) during adsorption and desorption stages, respectively. The adsorbent for an open

AWHA cycle with flow adsorber has to possess the adsorption sites with the affinity distributed in the ΔF-range between ΔF_{ad} and ΔF_{d}. In terms of the adsorption isotherm, it should be a gradual (or S-shaped) curve with uptake increasing between ΔF_{d} and ΔF_{ad} (Fig. 3.4, line 2). To raise the SWP and to minimize the necessary amount of the adsorbent the uptake variation Δw must be large. Such an adsorbent ensures high values of fractions δ_{ex}, δ_{col}, and RR, and promotes low SEC (Gordeeva et al. 2020). Based on climatic data for specific regions and the temperature of the heat source, available for the adsorbent regeneration, the values ΔF_{ad} and ΔF_{d} can be calculated as the quantitative requirements for the adsorbent properties for AWHA. Then, using these values and the characteristic curves of water adsorption on various materials, the appropriate adsorbent can be selected (Gordeeva et al. 2020; Solovyeva et al. 2021a). For a narrow temperature range, the adsorption isotherm $w(P/P_0)|_{T=\mathrm{cons}}$ and the data on RH-values of the ambient air during adsorption and desorption stages might be used to quickly assess the applicability of the adsorbent for AWHA under specific climatic conditions (Fig. 3.4). It should be noted that for various modifications of AWHA cycles and systems (e.g. semi-open cycles, or passive AWHA utilizing natural convection, etc.) the requirements for the adsorbent properties might be modified appropriately (see Sect. 3.3).

Considering the energy demands for water production, to decrease $\mathrm{SEC_{th}}$ (Eq. 3.3) and increase the thermal efficiency η_{th} (Eq. 3.7), the adsorbents with low desorption enthalpy ΔH_{d} have to be used. It should be mentioned, that assuming the density of the water adsorbed constant, the desorption enthalpy and adsorption potential are linked by the following expression (Dubinin and Stoeckli 1980; Andersson et al. 1985)

$$\Delta H_{\mathrm{d}}(w) = L + \Delta F(w) \tag{3.10}$$

Thus, the adsorbents with a strong affinity to water vapor ensure high efficiency of water adsorption from air δ_{ex}, but they require high thermal energy consumption

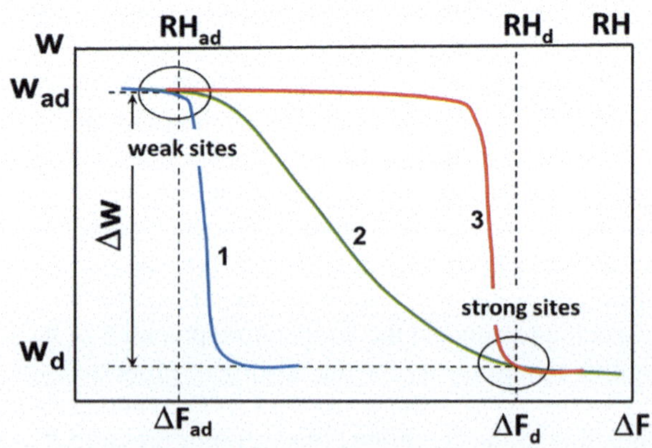

Fig. 3.4 The characteristic curve of water sorption of the adsorbent needed for a passive cycle (1), active open cycle with flow adsorber (2), and active semi-open cycle (3) of AWHA

SEC_{th} that results in lowering energy efficiency η_{th}. Thus, getting the right adsorbent selection involves a trade-off between these indexes.

3.3.1.2 Sorption Dynamics

The sorption dynamic behavior of the AWHA system depends on both material and component level properties (LaPotin et al. 2019; Yang et al. 2021). The detailed discussion of the component-level properties of the AWHA system is out of the scope of this chapter, so the dynamic issues will be described briefly. Since physical adsorption is a relatively fast process, the adsorption kinetics usually is controlled by mass and heat transfer resistances. The main resistances for mass transfer are intraparticle, surface, and interparticle diffusion resistance (Fig. 3.5a) (Ruthven 1984). For conventional pelleted adsorbents with the bimodal porous structure, the intraparticle diffusion can be controlled by diffusion in micropores and transport macro- or mesopores (Fig. 3.5b). The intraparticle mass transfer might occur through several mechanisms, among which there are the Knudsen, molecular, and surface diffusion (Yang et al. 2021; Ruthven 1984). Since adsorption is the exothermal process, heat transfer affects the sorption kinetics as well. Accordingly, three heat transfer resistances, namely, conduction in the adsorbent particle, and convection/radiation from the surface (Ruthven 1984). The relative contributions of all these resistances determine the adsorption kinetics, and an exact mathematical description of the process is a complicated task. For simplification purposes, the "linear driving force approximation" suggested by Glueckauf (1955) is often used for numerical simulation of the adsorption processes

$$\partial q/\partial t = 15/R_c^2 D\left(q_{eq} - q\right), \tag{3.11}$$

where R_c is the particle radius, D is the diffusivity, q and q_{eq} are the instantaneous and equilibrium vapor concentration in the adsorbent particle.

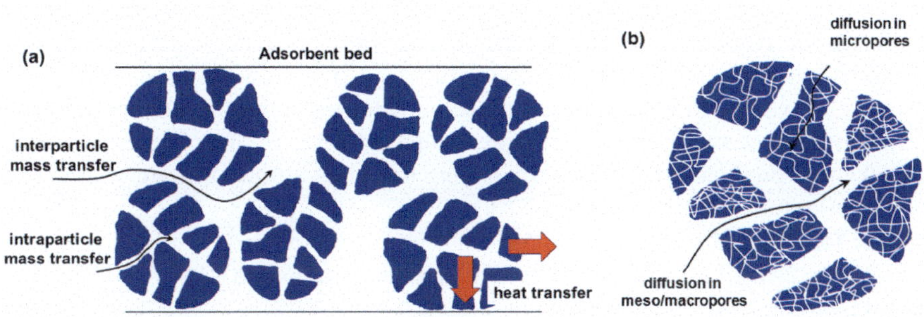

Fig. 3.5 Schematic presentation of different resistances affecting the adsorption kinetics (**a**) and intraparticle diffusion in micro- macro pores (**b**)

In addition to thermodynamic and dynamic issues, there are also general requirements, such as (i) hydrothermal stability, (ii) mechanical strength, (iii) no volatile impurities, which can pollute the collected water, and (iv) low price. Below the main classes of sorbent materials will be considered, keeping in mind these requirements.

3.3.2 Solid Adsorbents

3.3.2.1 Conventional Adsorbents

The conventional commercial adsorbents used for water vapor adsorption are microporous silica gels, activated alumina, and zeolites (Ng and Mintova 2008). Zeolites represent a wide family of microporous crystalline aluminosilicates, which framework is composed of $[SiO_4]$ and $[AlO_4]^-$ tetrahedra. Due to a negative framework charge, compensated by cations, the zeolites possess a strong affinity to water with the adsorption isotherms of I type according to the IUPAC classification (Thommes et al. 2015), with a large uptake at low P/P_0 from 0 to 0.03. Silica gels with the chemical composition $SiO_2 \cdot nH_2O$ are amorphous solids, which porous structure depends on the synthesis conditions and varies from micro- to mesoporous. Due to the electroneutral surface of silica gels, they demonstrate a moderate affinity to water vapor. The water adsorption equilibrium is described by isotherms of I type with a gradual uptake at a wide P/P_0 range for microporous silica gels. Water adsorption on mesoporous silica gels is described by isotherms of V type. The microporous silica of regular density (type RD) with pores of about 2.2 nm is usually used as a desiccant. Activated γ-alumina is characterized by a wide pore-size distribution in the mesopores range, with water vapor adsorption equilibrium described by isotherms of IV types.

A comparative study of the daily AWHA cycles using silica gel, activated alumina, and zeolite 13X as adsorbents and solar energy for water desorption was performed by Srivastava and Yadav (2018). The affinity of alumina to water is quite weak, it showed the lowest amount of water equal to 96 g/kg, and the minimum water production of 38 g/kg at a maximum desorption temperature of 150 °C. Zeolite with the strongest affinity to water vapor demonstrated good efficiency of water adsorption and capture 147 g/kg of water. However, due to its incomplete desorption at a temperature of 150 °C, it gave a small amount of water collected 43 g/kg. The highest SWP of 155 g/kg was produced with silica gel, having moderate affinity to water vapor, at a maximum desorption temperature of 100–110 °C with a primary energy efficiency of 18%. The SWP of silica beads in AWHA prototypes powered by solar heat varied from 105 to 241 g/kg at $RH_{ad} = 30–80\%$ and temperature 100–130 °C during the desorption (Das et al. 2022; Sleiti et al. 2021).

Gentile et all. (2022) showed that despite low water uptake on silica gel $\approx$0.1 g/g at RH = 30–50%, the lab-scale AWHA unit provided water at a low regeneration temperature of 57 °C with low specific solar energy consumption < 5 kWh/L and high efficiency of about 20–30%. Thus, the advantages of microporous silica gel are low regeneration temperature and specific energy consumption; its limitation is small specific water production, particularly at low RH of the ambient air. To increase the water adsorption capacity of silica, mesoporous silica nanofibers with enhanced adsorption ability were synthesized by electrospinning technique, which adsorbs up to 0.55–0.69 g/g. The growth of uptake was observed at a high RH range of 60–90%, so this adsorbent could be promising for AWHA in humid regions (Kima et al. 2019).

As shown above, conventional aluminosilicate zeolites possess a high affinity to water vapor and can be hardly used for AWHA powered by low-temperature solar heat. The family of crystalline aluminophosphates (AlPO) and silicon- or metal-substituted aluminophosphates (SAPO and MeAPO) are microporous crystalline solids with zeolite-like structures (Wilson et al. 1982). Their framework comprises Al–O and P–O tetrahedrons, (Martens and Jacobs 1994) and due to the electro-neutrality of their surface, they possess moderate variable hydrophilicity (Ng and Mintova 2008). The water adsorption equilibrium of AlPO, and MePO is described by stepped (or S-shaped) water adsorption isotherms. LaPotin et al. (2021) studied AWHA employing commercial microporous iron aluminophosphate AQSOA-Z01. Due to stepped water adsorption isotherm with sharp uptake at $P/P_0 = 0.15$–0.20 and shift of the step toward higher RH at increasing temperature, AQSOA Z01 can adsorb water at RH as low as 20% and desorb it at 60 °C if the condenser temperature equal to 25 °C. In the daily adsorption/desorption cycle 60 mL of water was collected from 520 g of the adsorbent using solar radiation for desorption. The specific water production SWP_S was estimated as 0.77 L/m^2/day.

3.3.2.2 MOFs

Recently, porous metal–organic frameworks (MOFs) have been considered an effective alternative to the currently used commercial water adsorbents (Yaghi et al. 2019). MOFs are a family of crystalline compounds that are commonly constructed by metal ions or metal clusters coordinated with polydentate organic ligands forming frameworks with ordered structures (Fig. 3.6) (Férey 2008; Yaghi et al. 2003; Kitagawa et al. 2004).

Among the existing porous solids, MOFs stand out by their unprecedentedly high porosity and specific surface area, large adsorption capacity, variable hydrophilicity, etc. (Yaghi et al. 2019). One of the most important advantages of MOFs is the possibility to control and rationally adjust their adsorption properties by modifying their structural and functional elements stemming from the principles of reticular chemistry (Yaghi et al. 2019, 2003; Canivet et al. 2014a). This has resulted in a huge variety of the porous structures of MOFs with diverse adsorption behavior and, as a consequence, different types of water adsorption isotherms (Canivet et al. 2014b). That makes it possible to select the MOFs with properties matching the climatic conditions of the specific region and endows MOFs with great potential for their application in AWHA from the air (Zhou et al. 2020; Tingting et al. 2020; Gordeeva et al. 2021).

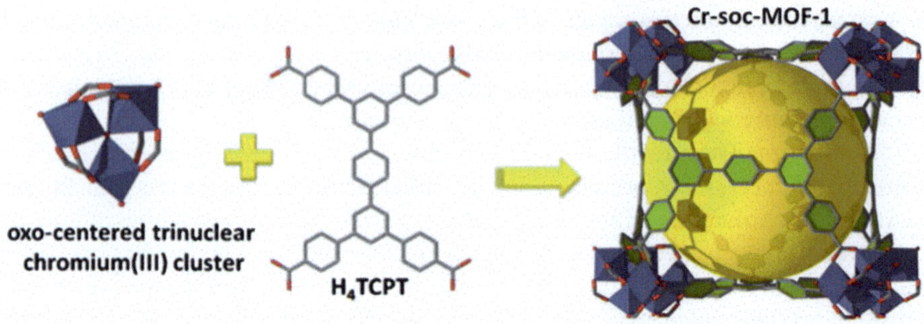

Fig. 3.6 The crystal structure representation of Cr-soc-MOF-1 consisting of the μ3-oxygen-centered trinuclear Cr(III) carboxylate clusters and the organic linker (3,3″,5,5″-tetrakis(4-carboxyphenyl)-*p*-terphenyl, H_4TCPT). Color code: C—gray; O—red; Cr—violet (Abtab et al. 2018)

The MIL-100 and MIL-101 series are highly stable and the most studied mesoporous MOFs. MIL-100(Fe) and MIL-101(Cr) were first proposed as advanced water adsorbents for potable water production in desert areas with a hot and dry climate by Seo et al. (2012). Later, Kim et al. (2016) estimated the water production of eight hydrothermally stable MOFs, namely, MIL-101(Cr) and the families of MIL-100(M) (M = Cr, Al, Fe) and UiO-66(Zr)-X (X = H, OH, $(OH)_2$, NH_2) under climatic conditions of three dry regions. It was demonstrated that MIL-101(Cr) and MIL-100(M) (M = Cr, Al, Fe) possess high water capacities of 0.35–1.09 g/g at the RH typical of the Atacama Desert in southern Peru (Pampas de La Joya) and the Mojave Desert in California. Noted that among these MOFs, MIL-101(Cr) and MIL-100(Fe) are currently produced on large scales at reasonable prices.

Recently, the AWHA potential of MIL-101(Cr), silica gel, and zeolite has been determined using Monte Carlo simulations coupled with climatic data for different areas of the United States (Mulchandani and Westerhoff 2020). The results predicted that the maximum $SDWP_s$ of 3.1 L/m^2/day can be achieved at RH ranging from 10 to 40% by using MIL-101(Cr) as a desiccant. Wherein, the total amount of water collected by the MOF, even with limited sunlight, can be up to double that of zeolite or silica in many regions. It should be noted that although the MIL-101(Cr) possesses a large water capacity (> 1.0 g/g) and excellent stability towards water (Akiyama et al. 2012), however, its affinity to water is weak resulting in a small uptake at a low RH < 50% typical of extra-dry climates. Introducing inorganic salts into the pores is an effective strategy to create additional adsorption centers and increase the affinity to water vapor. According to this approach, the composite sorbents were developed, described in more detail in Sect. 3.4.

Furukawa et al. (2014) have evaluated the water adsorption performances of 20 various MOFs. The MOF-801-P and MOF-841 were highlighted as adsorbents with the highest water capacity and water stability. The laboratory device for AWHA was designed employing MOF-801 as an adsorbent that gave SWP = 0.24 L/kg/cycle or SDWP = 2.8 L/kg/day (several cycles per day) at RH of about 20% and a temperature of 35 °C by only using low-grade heat from natural sunlight (1 kW/m^2) (Kim et al. 2017). It was predicted that this MOF allows obtaining the water amount of about 0.25 L/kg/day at RH of 10–40% in arid climates under outside conditions (Kim et al. 2018). Later, the water production by MOF-801 has been studied during field trials in the Arizona desert. The device equipped with kilogram quantities of MOF-801 delivers 0.2–0.3 L/kg per day at RH = 5–40% and ambient temperature of 20–40 °C (Xu and Yaghi 2020).

Recently, the concept of isoreticular chemistry has been successfully applied to develop aluminum-based MOF-303 which possesses improved hydrophilicity and chemical stability (Fathieh et al. 2018). It showed a considerable maximum water uptake of 0.48 kg/kg and was more cost-effective than MOF-801. It was shown that employing MOF-303 allowed the production of 1.3 L/kg_MOF/day in a laboratory environment (at RH = 32% and 27 °C) and 0.7 L/kg_MOF/day under outdoor conditions of the Mojave Desert (at RH = 10% and 27 °C) (Hanikel et al. 2019; Wasti et al. 2022; Logan et al. 2020).

Of particular interest is mesoporous Cr-soc-MOF-1 (Fig. 3.6), which shows high hydrolytic stability and enormous water uptake of 1.95 g/cm^3 at a relative pressure of 0.7, which far exceeds the appropriate values for other porous adsorbents (Abtab et al. 2018). Co_2Cl_2BTDD has one of the highest reported water adsorption capacities of 0.82 g/g under simulated desert conditions (5% RH at 45 °C in the day-time and 35% RH at 25 °C in the night-time) (Rieth et al. 2017). This is nearly double the previous best material MOF-841 reported for this application, which captures only 0.42 g/g of water. Although these MOFs have huge adsorption capacity, the complexity of the organic ligand synthesis and their high cost can significantly limit their further implementation for AWHA.

As noted earlier, the combination of diverse organic and inorganic building units of MOFs results in a virtually infinite number of possible structures (Tranchemontagne et al. 2009). To date, more than 100,000 different structures have already been registered in the Cambridge Structural Database (Moghadam et al. 2020). Thus, the screening methodology is needed to select the most promising MOFs for a specific application, among the many existing ones. A systematic approach to finding the optimal MOF for AWHA was suggested by Gordeeva et al. (2020). In the frame of this approach, the data on water adsorption equilibrium of the number of MOFs were presented as characteristic curves $w = f(\Delta F)$ (Fig. 3.7). The compliance of MOFs sorption behavior with the specific AWHA cycle requirements was analyzed taking into account the ΔF_{ad} and ΔF_d values corresponding the conditions of the adsorption and regeneration stages for selected climatic regions. According to this method, a number of the most promising MOFs were selected for AWHA in arid regions using solar energy (Fig. 3.7), and their efficiency was estimated in terms of fractions δ_{ex} of water extraction and δ_{col} of water collection and the specific mass Δw of the water exchanged in the cycle (Gordeeva et al. 2020). In particular, it was shown that MIL-101(Cr), MIL-101(Cr)-SO_3H, and Co_2Cl_2(BTDD) are appropriate for the Central Australian zone with a moderately humid climate. MIL-160 and CAU-10(pydc) are most suitable for the arid climates of Saudi Arabia and the Sahara Desert. In the considered climatic zones, these MOFs showed the specific capacity $\Delta w = 0.34$–1.6 g/g and allowed the fractions of water extraction $\delta_{ex} = 0.78$–0.93 and collection $\delta_{col} = 0.75$–0.90 to be achieved at the regeneration temperature of 75–100 °C.

The feasibility of AWEA employing MIL-160 under extremely arid climates of the Mojave and Sahara Deserts, Chile, Algeria, the central part of Saudi Arabia, and Mongolia was studied by Solovyeva et al. (2021a). Under the conditions of these regions, MIL-160 enabled the specific capacity $\Delta w = 0.31$–0.33 g/g per cycle at *RH* = 20% and high fractions $\delta_{ex} = 0.90$–0.98 and $\delta_{col} = 0.48$–0.97 at a regeneration temperature of 80–100 °C with the condenser cooled by ambient air. The specific energy consumption for water production was evaluated as 3.5–6.8 kJ/g, which is acceptable when using solar heat for regeneration. Besides, the application efficiency of MIL-160(Al) as an adsorbent for water harvesting was experimentally evaluated in a bench-scale fixed-bed unit (Silva et al. 2021). It was shown that MIL-160(Al) produced 305 L/ton/day of water at regeneration and condensation temperatures of 80 and 10 °C, respectively.

It is worth noting that despite the versatile advantages and potential of porous MOFs for the water harvesting process, one of the most common problems hindering the employment of many MOFs is their hydrothermal instability (Schoenecker et al. 2012; Kumar et al. 2019). At the same time, owing to the modern advances in the chemistry of MOFs, more and more materials with high water stability are developed (Wang et al. 2016). On the other hand, MOFs have been tailored mainly in lab settings, but the efficient production of large-scale batches makes them expensive and barely commercially available. Although the large-scale production of some MOF materials has been reported (Kalmutzki et al. 2018; Gaab et al. 2012) the prospect of their widespread usage is still dependent on growing global demand and investment.

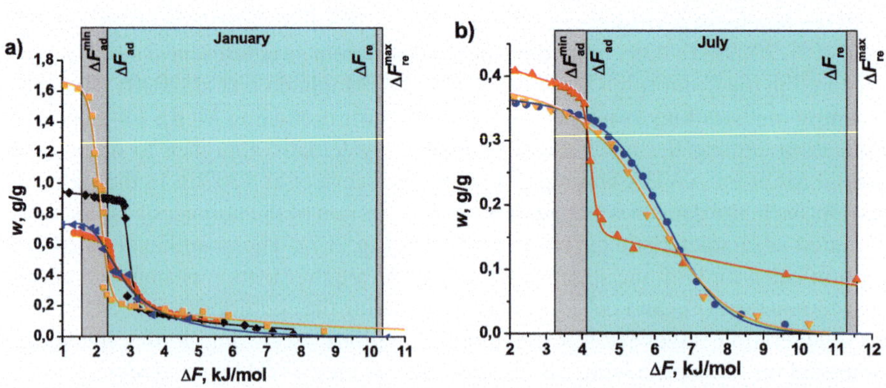

Fig. 3.7 Characteristic curves for water adsorption on **a** MIL-101(Cr) (orange filled square), Co_2Cl_2-BTDD (black diamond), MIL-100(Fe) (blue left pointing triangle), MIL-101(Cr)-SO_3H (red filled circle); **b** MIP-200 (red filled triangle), MIL-160, (blue filled circle, blue open circle), CAU-10(pydc) (orange filled inverted triangle). The values of ΔF_{ad}, $\Delta F_{ad}{}^{min}$, ΔF_{re}, and $\Delta F_{re}{}^{max}$ for the operating conditions adsorption and regeneration stages at $T_{re} = 80$ °C in the CA (**a**) and SA (**b**) zones [Reprinted from Gordeeva et al. (2020). Copyright 2019, with permission from Elsevier]

3.3.3 *Liquid and Solid Absorbents*

3.3.3.1 Salt Solutions

Among different liquid absorbents, an aqueous solution of $CaCl_2$, a widely available and cheap hygroscopic salt, is mainly used for AWHA. They possess high moisture absorbing capacity, the captured moisture can be released at a low regeneration temperature of 40–70 °C (Mohamed et al. 2017). AWHA systems employing salts solution as absorbents powered by solar thermal energy were widely studied both theoretically (Mohamed et al. 2017; William et al. 2013) and experimentally (Talaat et al. 2018; William et al. 2015) under climatic conditions of Al-Hada and Taif (Saudi Arabia) and Cairo (Egypt). The performance of the liquid absorbent bed was limited by slow mass transfer between the air and desiccant due to the small contact surface area. To increase the bed surface area inside the collector, the desiccant was mixed with a carrier (sand or cloth) (Mohamed et al. 2017; William et al. 2015). The main factors, affecting the system performance are the nature of the bed carrier, initial salt solution concentration, dew point, and relative humidity of the ambient air. Consequently, ab-/desorption surface was increased, which contributed to faster sorption. Although the theoretical specific water production of 3.0–3.1 L/m^2/day was predicted (Mohamed et al. 2017; William et al. 2013) the specific production in the range of 1–1.5 L/m^2/day was achieved experimentally at RH = 60–80%. The productivity decreased with an increase in the initial salt concentration of the solution. The solar energy efficiency for the cloth absorbent bed was equal to 29.3% which was higher than for the sandy bed (17.7%) (William et al. 2015) due to its high porosity and the solution soaking capacity.

The performance of AWHA based on liquid absorbents is mainly limited by slow vapor desorption. To accelerate water vapor desorption from the solution powered by solar heat, Wang et al. (2019) applied so-called interfacial solar heating to the AWHA unit, based on graphene oxide and cellulose composite aerogel placed on the surface of $CaCl_2$ solution (Fig. 3.8). Tailored GO-based aerogel with large macropores of hundreds of microns enabled fast water diffusion and capillary pumping of the salt solution toward the aerogel surface, thus ensuring fast water desorption. The SWP_s of 2.89 L/m^2/day was achieved in a daily cycle at 70% RH with solar energy efficiency as high as 66.9%.

3.3.3.2 Hydrogels

Hydrogels are a class of polymers, which can absorb or retain a large amount of water due to the presence of hydrophilic groups. The polymers are crosslinked by physical or chemical bonds forming a network (Zhang et al. 2022). Chemically crosslinked hydrogels are formed through covalent bonding that ensures better mechanical strength. The common routes for the synthesis of chemically bonded hydrogels are the free radical and radiation polymerization of monomers. During the

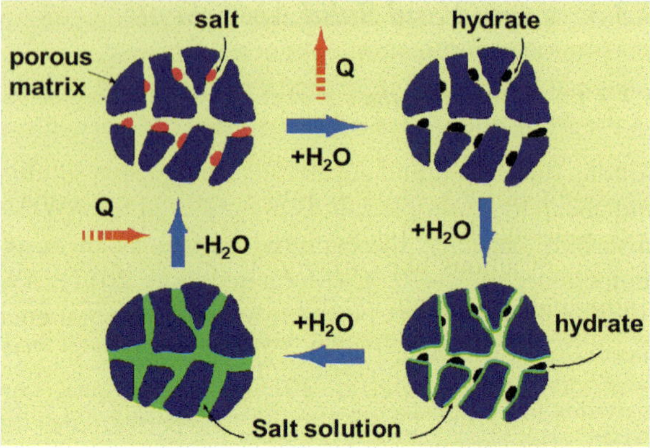

Fig. 3.8 Scheme of water sorption on CSPMs (Aristov 2003)

former, monomers containing a double bond (acrylic acid, acrylamide, vinylpyrrolidone) are used, and molecules with two or more double bonds capable of polymerization (*N*-methylene bisacrylamide, ethylene glycol acrylate) serve as cross-linking agents (Ida et al. 2021). During radical polymerization, saturated organic monomers can be used as monomers and cross-linking agents; the initial free radicals are generated by gamma or ultraviolet radiation, and electron beam irradiation (Moghaddam et al. 2019). Physically cross-linked hydrogels involve hydrogen, electrostatic, hydrophobic interactions, van der Waals force, and chain entanglement. Ionic bonds between polyelectrolytes, such as sodium alginate and polyacrylic acid, with opposite charges, and between polyelectrolytes and multi charged ions are formed in chemically bonded hydrogels (Wu et al. 2015). Amphiphilic polymers with both hydrophilic and hydrophobic moieties (e.g. dodecyl-modified polyacrylic acid, octadecyl acrylate) can form hydrogels crosslinked through an association between hydrophobic groups (Wei et al. 2018). There are also double network hydrogels, which represent two interpenetrated networks, one of which is chemically and another physically crosslinked (Xu et al. 2021). Due to the presence of highly hydrophilic polar groups ($–NH_2$ and $–SO_3$) and ionized groups ($–O^-$, $–COO^-$), hydrogels possess remarkable water swelling and absorption ability and offer new opportunities for AWHA.

Loo et al. (2020) reported hydrophilically enhanced photothermal foam (HEPF) that can generate potable water from seawater and atmospheric moisture via solar-powered evaporation at its interface. HEPF is a polymer network of polyurethane and highly hygroscopic poly(sodium acrylate) with embedded expanded graphite as solar absorbing material. HEPF provides 250–1770 g/g of water per cycle through moisture or liquid water absorption, followed by solar-driven evaporation and condensation. Zhao et al. (2019) developed super moisture absorbing gel (SMAG), composed of highly hygroscopic chloride-doped polypyrrole (PPy-Cl) and water storing poly-*N*-isopropylacrylamide (poly-NIPAM) gel with thermal-responsible hydrophilicity. Moisture is absorbed by PPy-Cl, liquefied, and then transferred into the poly-NIPAM network, causing its swelling. The water uptake on SMAG varies from 0.7 to 6.7 g/g at the RH range of 30–90%. Most of the stored humidity can be directly released as liquid water by heating up to 40 °C, which is enabled by a stimuli-responsive shrinking of poly-NIPAM. The rest of about 1.7 g/g can be thermally desorbed at ≈63 °C with further condensation. Thus, due to a synergistic combination of hygroscopic PPy-Cl, water swelling poly-NIPAM with thermally switched hydrophilicity, SMAG provided an SWP of 3.9 g/g at a relative humidity of 85%.

Despite the exciting swelling ability of hydrogels, their moisture absorbing capacity at low relative humidity typical of arid regions is limited. For this reason, AWHA employing hydrogels inspire high hopes for coast regions in high humidity circumstance.

3.3.4 Composite Sorbents Based on Hygroscopic Salts

Each of the sorbent classes considered above possesses both advantages, which might have the potential for improving AWHA systems, and drawbacks, limiting their implementation in actual practice. Thus, the adsorption capacity of common solid adsorbents with high hydrothermal stability (silica, zeolites) is quite small (0.1–0.25 kg/kg), resulting in low SWP of the systems. The adsorption capacity of several MOFs with extremely high porosity and specific surface area reaches impressive values of 1.6 and 1.95 kg/kg for MIL-101(Cr) (Canivet et al. 2014b) and Cr-Soc-MOF-1 (Abtab et al. 2018), respectively. However, their affinity to water vapor is quite weak and the maximum uptake is reached only at a high RH of 60–80% which limited their application for AWHA in arid regions, most severely affected by water scarcity. Furthermore, the low hydrothermal stability of MOFs can be also a challenge. The liquid absorbents (salt solutions) are able to capture large amounts of moisture at a wide range of RH making them promising for various climatic regions, however, their practical application is hindered by slow mass transport and crystallization of salt hydrates during desorption. Solid absorbents and hydrogels possess huge swelling rates up to thousands of their weight, but humidity absorbing ability at low RH is minor. For these reasons, a combination of benefits of various sorbent types in one physical–chemical structure of composite sorbents might promise great potential for AWHA. This approach was successfully applied for the development of absorbents SMAG (Zhao et al. 2019) and HEPF (Loo et al. 2020) based on the hydrogel. A similar approach was used in an atmospheric water generator, employing graphene oxide, cellulose composite aerogel, and $CaCl_2$ solution (Wang et al. 2019). Below, a large family of composites, based on hygroscopic salts embedded inside pores of solid matrixes, is considered.

3.3.4.1 Sorption Mechanisms

Composites "Salt inside Porous Matrix" (CSPMs) (Gordeeva and Aristov 2012) are composed of common porous adsorbents [silica gels (Aristov et al. 1996), alumina

(Gordeeva et al. 2000), activated carbons (Gordeeva et al. 2002), natural porous clays vermiculite (Casey et al. 2014), attapulgite (Hai-jun et al. 2008)], etc. and a hygroscopic salt [$CaCl_2$ (Daou et al. 2006), LiCl (Zhang et al. 2016), LiBr (Gordeeva et al. 1998a), MgSO4 (Posern et al. 2015), $MgCl_2$ (Yu et al. 2019), or their mixture (Gordeeva et al. 2013)], which is inserted inside the pores. The mechanism of water sorption on CSPM involves several processes (Fig. 3.8) (Gordeeva and Aristov 2012):

- at low P/P_0 water molecules are adsorbed on the active sites of the matrix surface;
- at increasing P/P_0 the salt S reacts with water according to the reaction

$$S + nH_2O \Leftrightarrow S \times nH_2O; \tag{3.12}$$

  a salt hydrate $S \times nH_2O$ (or hydrates with different number n of water molecules) forms, which deliquesces and transforms to an aqueous salt solution;
- finally, the concentrated salt solution absorbs water vapor giving a dilute solution. The salt solution is retained inside the matrix pores by capillary forces until its volume is smaller than the pore volume of the composite.

It was shown that the impact of adsorption on the matrix is usually minor for meso- and macroporous matrixes and uptake does not exceed 0.01–0.05 g/g. The salt is the main sorbing component, and the contribution of the reaction (3.12) and the absorption by the salt solution to the total uptake on the composite is dominant for both mesoporous (Gordeeva et al. 1998a, 2000) and macroporous (Korhammer et al. 2016) composites. The water uptake grows at the increasing salt content of the composite (Gordeeva and Aristov 2012; Gordeeva et al. 1998a; Permyakova et al. 2017). For composites, based on microporous matrixes [e.g. zeolites (Xu et al. 2019), aluminophosphates (Nguyen et al. 2020)], particularly those with a strong affinity to water vapor, the contribution of the matrix can be essential due to the volume filling of micropores at a low P/P_0. Xu et al. (2019), showed that the physisorption impact on the total sorption capacity of $MgCl_2$/zeolite 13X reaches 25.5%. However, it should be noted that microporous adsorbents (except some microporous MOFs) are characterized by a low specific pore volume usually not exceeding 0.2–0.4 cm^3/g, which strongly reduces the adsorption capacity of the microporous composite sorbents.

The solid porous matrix is of utmost importance as well. It supports the active salt, prevents the agglomeration of the salt particles, and retains the solution inside pores (Gordeeva and Aristov 2012). During ad/desorption, the matrix provides the mass transfer to/from the salt/hydrate particles through the pore system and ensures heat transfer through the matrix carcass, which promotes fast sorption. Furthermore, it was shown that the dispersion of the salt inside pores of the nanometer scale results in the alteration of their sorption properties (Gordeeva and Aristov 2012; Gordeeva et al. 2000, 2002). Thus, when inserted inside mesoporous matrixes with pore size > 10–15 nm, the salt is stabilized in the crystalline phase inside pores; during water sorption, the crystalline hydrates are formed. According to the Gibbs phase rule, the system is monovariant which is revealed as a step on the adsorption isotherms (Fig. 3.9a) (Grekova et al. 2018; Shkatulov et al. 2020). The deliquescence of the hydrates results in the formation of the solution with bivariant equilibrium with water

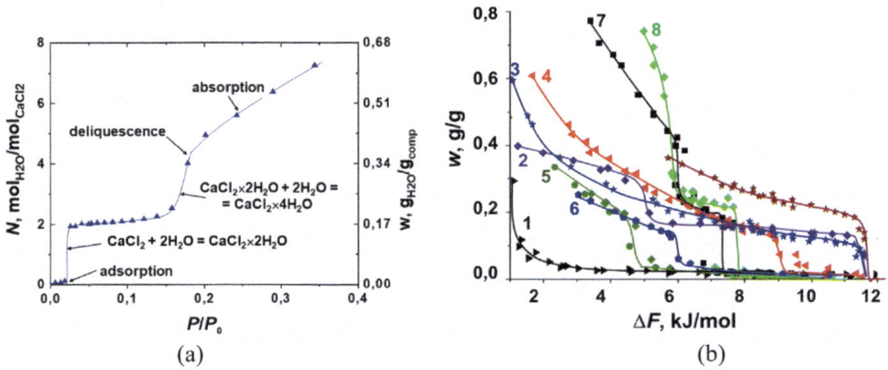

Fig. 3.9 Characteristic curves of water vapor sorption on CSPMs: **a** $CaCl_2$/silica (15 nm), **b** Na_2SO_4/silica (9 nm) (1), $MgSO_4$/silica (15 nm) (2), LiBr (32 wt%)/silica (15 nm) (3), LiCl/silica (15 nm) (4), $LiNO_3$/silica (15 nm) (5), $Ca(NO_3)_2$/silica (15 nm) (6), LiCl/MWCNT (7), LiCl/vermiculite (8)

vapor, and the uptake growths gradually at increasing P/P_0. In smaller pores of 2–6 nm size, the dispersed salt forms an X-ray amorphous phase. Such systems are bivariant with smooth sorption isotherms over the whole range of relative pressure from 0 to 1 (Shkatulov et al. 2020; Garzón-Tovar et al. 2017).

Due to the combination of three sorption mechanisms (physical adsorption on the matrix surface, chemical reaction between the salt and absorption by the salt solution) the CSPMs are characterized by a high water sorption capacity, exceeding 1 g/g (Garzón-Tovar et al. 2017; Grekova et al. 2016). Furthermore, smart selection of the salt, the matrix, and appropriate synthesis conditions provide powerful tools for the modification of the CSPM properties, or even tailoring the composites with the required sorption equilibrium (Fig. 3.9b), which matches the climatic conditions of a specific arid region (Gordeeva and Aristov 2012).

3.3.4.2 CSPMs for AWHA

Owing to high water sorption capacity and tunable sorption behavior SCPMs present exciting promise for AWHA and have attracted research interest for decades (Elmer and Hyde 1986; Alayli et al. 1987; Gordeeva et al. 1998b; Aristov et al. 1999). Inorganic salts LiCl (Wang et al. 2018, 2021a), $CaCl_2$ (Gordeeva et al. 1998b; Aristov et al. 1999; Srivastava and Yadav 2020), and LiBr (Gordeeva et al. 1998b), $MgCl_2$ (Zhao et al. 2021) characterized by high hygroscopicity and capable to react with water vapor at low RH of 1–10% are mainly used as hygroscopic salt. The use of the matrixes with a large pore volume is advantageous because it allows a larger amount of salt to be inserted, which contributes to the high water sorption capacity of CSPMs. Furthermore, such composites can hold more adsorbed water inside the pore space without leakage. Natural porous materials [expanded vermiculite (Kumar and Yadav

2017), coal ash (Srivastava and Yadav 2020), saw wood (Kumar and Yadav 2015)] can be used as the matrix for CSPMs. The composites based on $CaCl_2$ inside their pores, although exhibiting a moderate SWP of 0.092–0.195 L/kg per cycle, are very competitively priced.

Activated carbon fibers (ACF), characterized by a high specific surface area, porosity, and water-sucking capability, are often used as the carrier of the salt. The composite sorbent based on ACF impregnated with LiCl was studied for AWHA by Wang et al. (2018). Wide pose size distribution of the ACF with micro- meso- and macropores promoted fast mass transfer during sorption. The prototype composed of ca. 70 kg of the composite harvested 14.3–38.5 kg of water with RR = 41–54% at RH = 37–75% and desorption temperature of 90–93 °C. To easily shape the ACF and prevent their deformation after sorption, Wang et al. (2021a) developed a composite based on ACF felt stabilized with silica sol impregnated with LiCl. The adsorption capacity of the composites of 1.4 g/g was achieved at $P/P_0 = 0.6$. A forced air-cooled proof-of-concept water generator from island air with 21 kg of sorbent was constructed and tested. At 63% RH, and 31 °C the device generated up to 7.7 kg of water with RR = 33–70% and SEC_{th} of 6.2–7.7 MJ/kg.

Inspired by natural plant leaves a composite sorbent was developed by Wang et al. (Wang et al. 2022), which comprised a super-hydrophilic ACF matrix with high porosity of 92% loaded with LiCl and encapsulated by superhydrophobic fibrous skin. Elastic skin allowed efficient vapor transport to/from the hydrophilic core, accommodated the swelling of LiCl/ACF core during sorption, and prevented the formed solution from leakage. Owing to the large porosity of ACF matrixed, the high hygroscopicity of LiCl, and its ultra-high loading of 93 wt%, the composite sorbed about 2.2 g/g at 36% RH. In outdoor experiments in Hong Kong, a water harvester of 15 cm diameter powered by natural sunlight demonstrated the specific water production of 2.37 g/g/day at 60–70% RH.

One of the promising strategies is combining a hygroscopic salt as water capturing component, a matrix with high porosity and hierarchical pore structure as a carrier, and a carbonaceous material as solar heat absorbing material. Wang et al. (2021b) suggested nanostructured biopolymer hygroscopic aerogels (NBHA) comprising LiCl, nanofibrillated cellulose hydrophilic, skeleton and a graphene solar absorber, which demonstrated the equilibrium water uptake of 0.55–0.95 g/g at RH = 18–42%. Outdoor testing of the NBHA-based AWHA device powered by solar heat showed that the composite sorbed up to 1.46 g/g at RH ranging from 58.7 to 76.6%, and produced about 0.36 g/g or 0.42 kg/m^2 of water per day at a solar flux of 0.1–0.56 kW/m^2 (the city of Harbin).

MOFs with unprecedented porosity, high pore volume up to 2 cm^3/g, and uniform micro- mesopores have aroused considerable interest as matrixes for salts. The composite sorbent based on hygroscopic salt $CaCl_2$ inside pores of MIL-101(Cr) was developed for AWHA technology in extremely arid climate conditions of the Sahara Desert and Saudi Arabia regions. It was shown that the $CaCl_2$/MIL-101 composite with the salt content of 29% demonstrated the specific water capacity of 0.52–0.73 g/g per cycle under these conditions and provides $\delta_{ex} = 0.95–0.98$ and

$\delta_{col} = 0.61–0.97$ at regeneration temperature of 80–100 °C with the specific energy consumption of 2.9–4.4 kJ/kg (Solovyeva et al. 2021b).

The composite sorbent based on LiCl encapsulated inside MIL-101(Cr) was developed for efficient water harvesting from arid air by Xu et al. (2020). It was reported that the composite harvested about 0.77 g/g of water under an arid climate at 30% RH and 30 °C. In addition, the AWHA lab-scale prototype employing this sorbent was designed, which allowed the collection of 0.7 kg/kg under laboratory conditions and 0.45 kg/kg under ambient conditions using natural sunlight (0.5–0.8 kW m^{-2}). Also noted is that since the price of the Cr-based MOF is relatively cheap, the composite sorbent shows promising potential for large-scale AWHA applications.

Also, LiCl salt has been introduced into a HKUST-1 to create composite desiccant material that possesses the water uptake of 1.09 g/g and 0.5 g/g at 50% and 30% RH, respectively, and temperature of 25 °C (Zhao et al. 2020; Gado et al. 2022). The utilization of these sorbents can be limited by the low hydrothermal stability of HKUST-1. $CaCl_2$ encapsulated inside Fe-based ferrocenyl MOF with a porous hollow structure and excellent photo-thermal conversion ability was described by Hu et al. (2021). The water was sorbed by $CaCl_2$ (2.68 g/g_CaCl2 at 80% RH), and released under the solar light due to photothermal conversion enabled by Fe-FC-HCP.

Hydrogels are another class of solids, which might offer exciting opportunities for developing CSPMs for AWHA. Due to their impressive swelling ability hydrogels are able to retain a huge amount of water. The introduction of hygroscopic salts inside hydrogels makes it possible to obtain the composites capable of effective vapor sorption and holding the formed solution inside the gel. The hybrid sorbent, comprising polyacrylamide and alginate crosslinked double network hydrogel, loaded with deliquescent salt ($CaCl_2$) and carbon nanofibers as a photothermal agent, was developed by Park et al. (2022) for AWHA (Fig. 3.10). A prototype employing the rapid adsorption–desorption ratcheting instead of common daily cycles with the composite gel enabled the production of 1.81 g/g of water per day at 66% RH.

A composite, prepared by grafting thermo-responsive polymer Poly(*N*-isopropylacrylamide) (PNIPAM) to silica gel and impregnated with LiCl to enhance the affinity to water vapor was developed (Ma and Zheng 2022). The adsorption capacity of the composite reached 1.7 g/g at 20 °C and 70% RH. Thermo-responsive

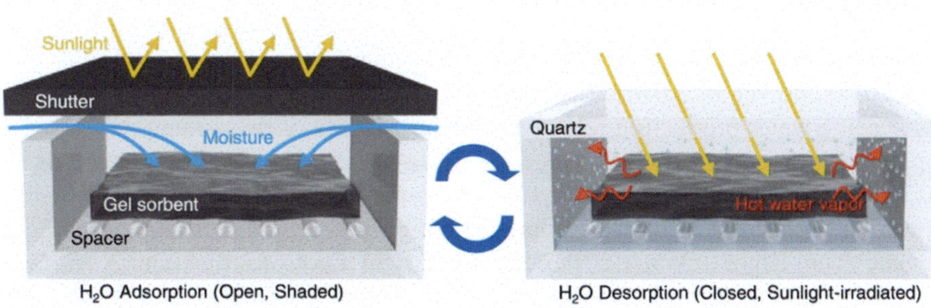

Fig. 3.10 Schematic illustrations of adsorption, desorption, and condensation in a prototype condensation chamber (Park et al. 2022)

PNIPAM allowed 56% of water can be released at an extremely low desorption temperature of 40 °C at 10% RH. A flexible hybrid photothermal sorbent composed of $CaCl_2$ embedded into polyacrylamide gel and carbon nanotubes as a photothermal component was developed in Li et al. (2018). A prototype delivered 0.52 g/g at 60–70% RH.

Thus, the combination in one physical–chemical structure of various sorbents, the advantages of which, such as high hygroscopicity of inorganic salts, high porosity of MOFs and ACF, the enormous water retention ability of hydrogels, can be synergistically enhanced, opens up unprecedented opportunities for the development of effective composite sorbents for AWHA. The performance of actual AWHA devices also depends on the components and system configuration and cycle management. Below various cycles and system configurations are briefly considered.

3.4 Cycles and Configurations

3.4.1 Active Versus Passive Cycles

According to energy consumption, AWHA can be divided into passive and active systems. The passive systems can be powered entirely by natural or sustainable energy without the consumption of electricity or other high-grade power for vapor desorption or forced convection, and spontaneously harvest water in the air under different atmospheric conditions (Jarimi et al. 2020). The adsorption stage is realized by exposing the adsorbent layer to the ambient air without forced convection (Fig. 3.10), while during the desorption stage the adsorber is closed by a transparent cover so that the adsorbent is heated directly by sunlight to enable the vapor desorption and condensation on the cold surface. The fundamental advantage of these systems is the ability to use environmentally friendly and free heat sources such as solar energy (Wang et al. 2017a), biomass (Chaitanya et al. 2018), ground heat (Heidarinejad et al. 2020) or waste heat (Vidhi 2018). The passive AWHA operates on a single daily cycle and extracts water at night and condenses during the day using usually solar thermal energy (LaPotin et al. 2021; Kim et al. 2018; Fathieh et al. 2018; Liu et al. 2022). Although the passive AWHA systems are beneficial due to their low energy consumption, water yield is significantly affected by climatic conditions (Tu et al. 2018; Gido et al. 2016; Jarimi et al. 2020; Lord et al. 2021). Furthermore, the low specific water production related to the unit surface of the sorbent restricts the implementation of passive cycles for large-scale AWHA units. Indeed, according to the World Health Organization, 50–100 L of water per person per day is needed to ensure the most basic needs, including drinking, food preparation, sanitation, and hygiene (The) Right to Water 2010). For a family of 4 persons, the required surface of the adsorbent bed can be estimated as 100–200 m^2 considering the SWPs of 1 L/m^2/day.

The employment of an active AWHA device (Fig. 3.2a) through the application of auxiliary systems for the adsorber purge, heating, and vapor condensation, allows the increase in daily water production and overcomes restrictions of climatic and geographical conditions. Active AWHA devices are not limited by a single daily cycle and can operate in continuous mode by multiple cycles per day (Hanikel et al. 2019; LaPotin et al. 2021; Lord et al. 2021). At the same time, it should be noted that the water yield of active systems is related to electric energy consumption which can lead to significantly higher costs (Lord et al. 2021).

It is worth noting that the requirements of the passive cycles to the adsorbent properties can differ significantly from those outlined above for active cycles with flow granulated adsorbent bed. Because of no energy consumption for the air blowing through the adsorber in the passive cycles, the significance of the fraction δ_{ex} of water extraction is minor. Accordingly, the strong adsorption sites, which adsorb water at a low RH and enable effective water extraction from the airflow and its dehumidification, are not necessary. The adsorbent with a quite low affinity, which captures a large amount of water under conditions of the adsorption stage and can be regenerated at a low temperature T_d, enabling a high fraction δ_{col}, can be applied. In other words, such an adsorbent possesses a high uptake $w(\Delta F_{ad})$ under conditions of the adsorption stage (Fig. 3.4, line 1) and is characterized by stepped adsorption isotherm with the step at adsorption potential equal to (or somewhat higher) ΔF_{ad}. Aluminophosphate AQSOA-Z01 (LaPotin et al. 2021), MOFs Co_2Cl_2-BTDD (Rieth et al. 2017), MIL-101(Cr) (Seo et al. 2012) and several CSPMs based on LiCl (Wang et al. 2018, 2021a) might be advantageous for passive cycles for various climatic conditions.

3.4.2 *Open Versus Semi-open Systems*

Currently, two types of AWHA systems have been suggested: open (Aristov et al. 1999) and semi-open (Wang et al. 2017b). In the open AWHA systems, the airflow both after passing through the adsorbent bed during the adsorption stage and after condensing the water during the regeneration stages is released into the environment. On the contrary, in the semi-open type devices the air employed for the sorbent regeneration continuously circulates in the system (Wang et al. 2017a, b).

In general, the system is open-type during sorption and closed-type during desorption and can be described by the following working phases (Fig. 3.11):

1. *Adsorption open phase*: The ambient air enters the adsorber, and the water vapor in the air is captured by the adsorbent. The sorption heat is released into the ambient air and the dry air flows out of the system (Fig. 3.11).
2. *Desorption and condensation closed phase*: The process air is heated and then flows to the adsorber. As a result, it is heated and the water vapor is desorbed to the air. The humid air flows to the condenser, where water is condensed into liquid water and then flows to the collector. The process air circulates in the

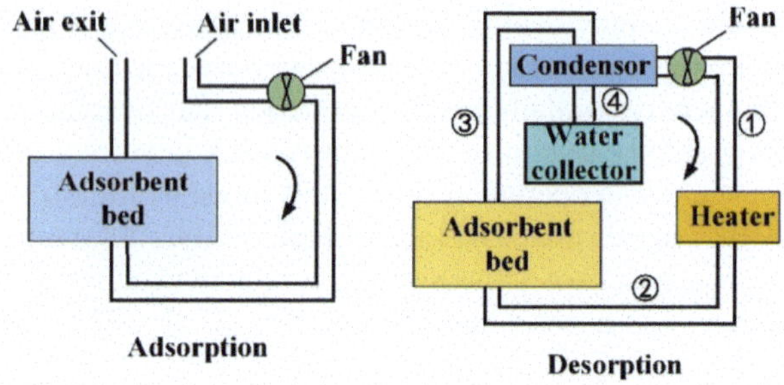

Fig. 3.11 The working scheme of the semi-open type of AWHA systems: (left) Sorption phase; (right) Desorption and condensation closed phase [Reprinted from Wang et al. (2017b). Copyright 2017, with permission from Elsevier]

device by a fan, and after the condenser, it goes through the adsorber bed again for the continuous desorption and condensation stage (Fig. 3.11).

Owing to the process air during desorption/condensation stages circulated between the adsorber and condenser, the vapor remaining in the process air after condensation is not rejected to the ambient, but reenters to the adsorber to absorb the desorbed water. This results in the increase in the fraction δ_{col} of water collected in the semi-open AWHA cycle. Accordingly, the weak adsorption sites, which promote efficient water desorption/collection become less important. The adsorbent with a quite strong affinity to water vapor, which captures/releases water at the adsorption potential equal to (or somewhat lover) ΔF_{d}, might be advantageous (Fig. 3.4, line 3). Such an adsorbent, on the one hand, enables effective water extraction (high fraction δ_{ex}) during the open adsorption stage. On the other hand, it can be regenerated under conditions of the close desorption stage at ΔF_{d} with a high fraction δ_{col}.

3.5 Summary and Outlooks

Owing to the wide availability of atmospheric moisture around the world, the ability to harvest water regardless of the climatic conditions and geographical location of the region, and the possibility to use solar and waste heat as driving energy sources, AWHA presents exciting prospects for decentralized potable water supply in the arid and remote of coastline areas, for emergency water supply after a natural and humanitarian disaster, etc. Nowadays, the feasibility of AWHA technology has been demonstrated by numerous lab-scale devices. The desiccant properties, and particularly their agreement with the climatic conditions of the specific region and the working cycle, are the cornerstones of potentially revolutionary advancements in this field. To date, a huge number of novel sorbents with advanced properties have been developed, which can be rationally designed to meet the requirements of a particular cycle. Among them, there are MOFs, hydrogels, etc. A promising strategy is a combination of the hygroscopic salts as a water-capturing matter, porous matrixes as the carrier for the salt and its solution providing fast mass- and heat transfer, and solar heat absorbing materials, into a single physical–chemical structure of the composite sorbents. However, there is still a large room for improving the properties of the employed sorbents for scaling up this technology. In our opinion, the main efforts of the current research on the AWHA are focused on:

- search for or rational design of novel advanced sorbents with a large specific uptake swing under conditions of regions with an extra-arid climate, most vulnerable to water scarcity;
- development of composite sorbents based on hygroscopic salts that provides improved mass transfer and prevents the liquid solution from leakage;
- synthesis of high porous MOFs with improved hydrothermal stability and enhanced affinity to water vapor;
- development of novel cycles, such as passive cycles without external electric or high-grade power supply, semi-open cycles with improved desorption/condensation efficiency, and cycles with sorption/desorption ratcheting.

We hope that advances in material science and applied engineering will allow these challenges to be addressed which will promote the further spread of AWHA technology.

Acknowledgements This work was supported by the Ministry of Science and Higher Education of the Russian Federation within the governmental order for Boreskov Institute of Catalysis (project AAAA-A21-121011390006-0).

References

Abtab SMT, Alezi D, Bhatt PM, Shkurenko A, Belmabkhout Y, Aggarwal H, Weselinski LJ, Alsadun N, Samin U, Hedhili MN, Eddaoudi M (2018) Reticular chemistry in action: a hydrolytically stable MOF capturing twice its weight in adsorbed water. Chem 4:94–105. https://doi.org/10.1016/j.chempr.2017.11.005

Akiyama G, Matsuda R, Sato H, Hori A, Takata M, Kitagawa S (2012) Effect of functional groups in MIL-101 on water sorption behavior. Micropor Mesopor Mater 157:89–93. https://doi.org/10.1016/j.micromeso.2012.01.015

Alayli Y, Hadji NE, Leblond J (1987) A new process for extraction of water from air. Desalination 67:227–229. https://doi.org/10.1016/0011-9164(87)90246-3

Altenkirch E (1934) Method for gaining water out of the atmosphere. US patent 2,138,689, 29 Nov 1934

Andersson JY, Bjurstrom H, Azoulay M, Carlsson B (1985) Experimental and theoretical investigation of the kinetics of the sorption of water vapour by silica gel. Farad Trans i, A J Phys Chem 11:2681. https://doi.org/10.1039/F19858102681

Aristov YuI, Tokarev MM, Cacciola G, Restuccia G (1996) Selective water sorbents for multiple application, 1. $CaCl_2$ confined in mesopores of silica gel: sorption properties. React Kinet CatAl Lett 59:325–335. https://doi.org/10.1007/BF02068130

Aristov YI (2003) Selective water sorbents: a new family of materials for sorption cooling/heating: state-of-the art. In: Proceedings of V Minsk international seminar on heat pipes, heat pumps, and refrigerators, Minsk, Belarus, 8–11 Sep 2003

Aristov YI, Tokarev MM, Gordeeva LG, Snytnikov VN, Parmon VN (1999) New composite sorbents for solar-driven technology of fresh water production from the atmosphere. Sol Energy 66(2):165–168. https://doi.org/10.1016/S0038-092X(98)00110-8

Asim N, Badiei M, Alghoul MA, Mohammad M, Samsudin NA, Amin N, Sopian K (2021) Sorbent-based air water-harvesting systems: progress, limitation, and consideration. Rev Environ Sci Biotechnol 20:257–279. https://doi.org/10.1007/s11157-020-09558-6

Bering BP, Dubinin MM, Serpinsky VV (1966) Theory of volume filling for vapor adsorption. J Colloid Interface Sci 21:378. https://doi.org/10.1016/0095-8522(66)90004-3

Beysens D, Milimouk I (2000) Pour les resources alternatives en eau (The case for alternative fresh water sources). Sécheresse 11:281–288

Canivet J, Bonnefoy J, Daniel C, Legrand A, Coasne B, Farrusseng D (2014a) Structure-property relationships of water adsorption in metaleorganic frameworks. New J Chem 38:3102–3112. https://doi.org/10.1039/C4NJ00076E

Canivet J, Fateeva A, Guo Y, Coasne B, Farrusseng D (2014b) Water adsorption in MOFs: fundamentals and applications. Chem Soc Rev 43:5594–5617. https://doi.org/10.1039/C4CS00078A

Casey SP, Elvins J, Riffat S (2014) Robinson A. Salt impregnated desiccant matrices for 'open' thermochemical energy storage—selection, synthesis and characterization of candidate materials. Energy Build 84:412–425. https://doi.org/10.1016/j.enbuild.2014.08.028

Chaitanya B, Bahadur V, Thakur AD, Raj R (2018) Biomass-gasification-based atmospheric water harvesting in India. Energy 165:610–621. https://doi.org/10.1016/j.energy.2018.09.183

Daou K, Wang RZ, Xia ZZ (2006) Development of a new synthesized adsorbent for refrigeration and air conditioning applications. Appl Therm Engn 26:56–65. https://doi.org/10.1016/j.applthermaleng.2005.04.024

Das A, Sharma R, Thirunavukkarasu V, Cheralathan M (2022) Desiccant-based water production from humid air using concentrated solar energy. J Therm Anal Calorim 147:2641–2651. https://doi.org/10.1007/s10973-021-10558-z

de Lange MF, Verouden KJFM, Vlugt TJH, Gascon J, Kapteijn F (2015) Adsorption-driven heat pumps: the potential of metal-organic frameworks. Chem Rev 115:12205–12250. https://doi.org/10.1021/acs.chemrev.5b00059

Domen JK, Stringfellow WT, Camarillo MK, Gulati S (2014) Fog water as an alternative and sustainable water resource. Clean Tech Environ Policy 16:235–249. https://doi.org/10.1007/s10098-013-0645-z

Dubinin MM, Stoeckli HF (1980) Homogeneous and heterogeneous micropore structures in carbonaceous adsorbents. J Colloid Interface Sci 75:34. https://doi.org/10.1016/0021-9797(80)90346-X

Elmer TH, Hyde JF (1986) Recovery of water from atmospheric air in arid Climates. Sep Sci Technol 21(3):251–266. https://doi.org/10.1080/01496398608058376

Fathieh F, Kalmutzki MJ, Kapustin EA, Waller PJ, Yang J, Yaghi OM (2018) Practical water production from desert air. Sci Adv 4:eaat3198. https://doi.org/10.1126/sciadv.aat319

Férey G (2008) Hybrid porous solids: Past, present, future. Chem Soc Rev 37:191–214

Furukawa H, Gándara F, Zhang Y-B, Jiang J, Queen WL, Hudson MR, Yaghi OM (2014) Water adsorption in porous metal–organic frameworks and related materials. J Am Chem Soc 136(11):4369–4381. https://doi.org/10.1021/ja500330a

Gaab M, Trukhan N, Maurer S, Gummaraju R, Müller U (2012) The progression of Al-based metal-organic frameworks-From academic research to industrial production and applications. Micropor Mesopor Mater 157:131–136. https://doi.org/10.1016/j.micromeso.2011.08.016

Gado MG, Nasser M, Hassan AA, Hassan H (2022) Adsorption-based atmospheric water harvesting powered by solar energy: comprehensive review on desiccant materials and systems. Process Saf Environ Prot 160:166–183. https://doi.org/10.1016/j.psep.2022.01.061

Garzón-Tovar L, Pérez-Carvajal J, Imaz I, Maspoch D (2017) Composite salt in porous metal-organic frameworks for adsorption heat transformation. Adv Funct Mater 27:1606424. https://doi.org/10.1002/adfm.201606424

Gentile V, Bozlar M, Meggers F, Simonetti M (2022) Liter-scale atmospheric water harvesting for dry climates driven by low temperature solar heat. Energy 254:124295. https://doi.org/10.1016/j.energy.2022.124295

Gido B, Friedler E, Broday DM (2016) Assessment of atmospheric moisture harvesting by direct cooling. Atmos Res 182:156–162. https://doi.org/10.1016/j.atmosres.2016.07.029

Glueckauf E (1955) Trans Faraday Soc 51:1540–1551. https://doi.org/10.1039/TF9555101540

Gordeeva LG, Aristov YI (2012) Composites "salt inside porous matrix" for adsorption heat transformation: a current state of the art and new trends. Int J Low Carbon Technol 7(4):288–302. https://doi.org/10.1093/ijlct/cts050

Gordeeva LG, Restuccia G, Cacciola G, Aristov YI (1998a) Selective water sorbents for multiple applications: 5. LiBr confined in mesopores of silica gel: sorption properties. React Kinet Catal Lett 63:81–88. https://doi.org/10.1007/BF02475434

Gordeeva LG, Tokarev MM, Parmon BN, Aristov YI (1998b) Selective water sorbents for multiple applications: 6. Fresh water production from the atmosphere. React Kinet Catal Lett 65:153–160. https://doi.org/10.1007/BF02475329

Gordeeva LG, Restuccia G, Tokarev MM, Cacciola G, Aristov YI (2000) Adsorption properties of the system "Lithium bromide—water" confined to pores of Expanded graphite, Sibunit and alumina. Rus J Phys Chem 74(11):2065–2069

Gordeeva LG, Freni A, Restuccia G, Aristov YI (2002) Water sorption on composites "LiBr in porous carbons." Fuel Process Technol 79(3):225–231. https://doi.org/10.1016/S0378-3820(02)00186-8

Gordeeva LG, Freni A, Restuccia G, Aristov YI (2007) Influence of characteristics of methanol sorbents "salts in mesoporous silica" on the performance of adsorptive air conditioning cycle. Ind Eng Chem Res 46:2747–2752. https://doi.org/10.1021/ie060666n

Gordeeva LG, Freni A, Krieger TA, Restuccia G, Aristov YI (2008) Composites "lithium halides in silica gel pores": methanol sorption equilibrium. Micropor Mesopor Mater 112:254–261. https://doi.org/10.1016/j.micromeso.2007.09.040

Gordeeva L, Grekova A, Krieger T, Aristov Y (2013) Composites "binary salts in porous matrix" for adsorption heat transformation. Appl Therm Eng 50:1633–1638. https://doi.org/10.1016/j.applthermaleng.2011.07.040

Gordeeva LG, Solovyeva MV, Sapienza A, Aristov YI (2020) Potable water extraction from the atmosphere: potential of MOFs. Renew Energy 148:72–80. https://doi.org/10.1016/j.renene.2019.12.003

Gordeeva LG, Tu YD, Pan Q, Palash ML, Saha BB, Aristov YI, Wang RZ (2021) Metal-organic frameworks for energy conversion and water harvesting: a bridge between thermal engineering and material science. Nano Energy 84:105946. https://doi.org/10.1016/j.nanoen.2021.105946

Grekova D, Gordeeva LG, Aristov YI (2016) Composites "Li/Ca halogenides inside multi-wall carbon Nano-tubes" for adsorptive heat storage. Sol Energy Mater Sol Cells 155:176–183. https://doi.org/10.1016/j.solmat.2016.06.006

Grekova A, Gordeeva L, Lu Z, Wang R, Aristov Y (2018) Composite "LiCl/MWCNT" as advanced water sorbent for thermal energy storage: sorption dynamics. Sol Energy Mater Sol Cells 176:273–279. https://doi.org/10.1016/j.solmat.2017.12.011

Griffiths JF (ed) (1972) World survey of climatology: climates of Africa, vol 10. Elsevier, Amsterdam, London, New York; Takahashi K, Arakawa H (eds) (1981) World survey of climatology: climates of Southern and Western Asia, vol 9. Elsevier, Amsterdam, Oxford, New York; Arakawa H (ed) (1969) World survey of climatology: climates of Northern and Eastern Asia, vol 8. Elsevier, Amsterdam, London, New York; Gentilli J (ed) (1971) World survey of climatology: climates of Australia and New Zealand, vol 9. Elsevier, Amsterdam, London, New York

Hai-jun C, Qun C, Ying T, Xiu-jun C, Hu-qing Y (2008) Attapulgite based LiCl composite adsorbents for cooling and air conditioning applications. Appl Therm Eng 28:2187–2193. https://doi.org/10.1016/j.applthermaleng.2007.12.015

Hanikel N, Prevot MS, Fathieh F, Kapustin EA, Lyu H, Wang H, Diercks NJ, Glover TG, Yaghi OM (2019) Rapid cycling and exceptional yield in a metal-organic framework water harvester. ACS Cent Sci 5:1699–1706. https://doi.org/10.1021/acscentsci.9b00745

Heidarinejad G, Rayegan S, Pasdarshahri H (2020) Dynamic simulation of a solar desiccant cooling system combined with a ground source heat exchanger in humid climates. J Build Eng 28:101048. https://doi.org/10.1016/j.jobe.2019.101048

Hu Y, Fang Z, Ma X, Wan X, Wang S, Fan S, Ye Z, Peng X (2021) CaCl2 Nanocrystals decorated photothermal Fe-ferrocene MOFs hollow microspheres for atmospheric water harvesting. Appl Mater Today 23:101076. https://doi.org/10.1016/j.apmt.2021.101076

Ibrahim NI, Al-Sulaiman FA, Ani FN (2018) Solar absorption systems with integrated absorption energy storage—a review. Renew Sustain Energy Rev 82:1602–1610. https://doi.org/10.1016/j.rser.2017.07.005

Ida S, Toda S, Oyama M, Takeshita H, Kanaoka S (2021) Multiarm star-crosslinked hydrogel: polymer network with thermoresponsive free-end chains densely connected to crosslinking points. Macromol Rapid Commun 42:2000558. https://doi.org/10.1002/marc.202000558

Jarimi H, Powell R, Riffat S (2020) Review of sustainable methods for atmospheric water harvesting. Int J Low-Carbon Technol 15:253. https://doi.org/10.1093/ijlct/ctz072

Kalmutzki MJ, Diercks CS, Yaghi OM (2018) Metal-organic frameworks for water harvesting from air. Adv Mater 30(37):1704304. https://doi.org/10.1126/science.aam8743

Khalil B, Adamowski J, Shabbir A, Jang C, Rojas M, Reilly K, Ozga-Zielinski B (2016) A review: dew water collection from radiative passive collectors to recent developments of active collectors. Sustain Water Resour Manag 2:71–86. https://doi.org/10.1007/s40899-015-0038-z

Kim SI, Yoon TU, Kim MB, Lee SJ, Hwang YK, Chang JS, Kim HJ, Lee HN, Lee UH, Bae YS (2016) Metal-organic frameworks with high working capacities and cyclic hydrothermal stabilities for fresh water production. Chem Eng J 286:467–475. https://doi.org/10.1016/j.cej.2015.10.098

Kim H, Yang S, Rao SR, Narayanan S, Kapustin EA, Furukawa H, Umans AS, Yaghi OM, Wang EN (2017) Water harvesting from air with metal-organic frameworks powered by natural sunlight. Science 356:430–434. https://doi.org/10.1126/science.aam8743

Kim H, Rao SR, Kapustin EA, Zhao L, Yang S, Yaghi OM, Wang EN (2018) Adsorption-based atmospheric water harvesting device for arid climates. Nat Commun 9:1–8. https://doi.org/10.1038/s41467-018-03162-7

Kima S, Park H, Choi H (2019) Maneuvering the ordered mesoporosity of electrospun silica nanofibers for water harvesting. Micropor Mesopor Mater 28:23–31. https://doi.org/10.1016/j.micromeso.2019.02.037

Kitagawa S, Kitaura R, Noro S (2004) Functional porous coordination polymers. Angew Chem Int Ed 43:2334–2375. https://doi.org/10.1002/anie.200300610

Klemm O, Schemenauer RS, Lummerich A, Cereceda P, Marzol V, Corell D, Van Heerden J, Reinhard D, Gherezghiher T, Olivier J et al (2012) Fog as a fresh-water resource: overview and perspectives. Ambio 41:221–234. https://doi.org/10.1007/s13280-012-0247-8

Kogan B, Trahtman A (2003) The moisture from the air as water resource in arid region: hopes, doubts and facts. J Arid Environ 53:231–240. https://doi.org/10.1006/jare.2002.1028

Korhammer K, Druske MM, Fopah-Lele A, Rammelberg HU, Wegscheider N, Opel O, Osterland T, Ruck W (2016) Sorption and thermal characterization of composite materials based on chlorides for thermal energy storage. Appl Energy 162:1462–1472. https://doi.org/10.1016/j.apenergy.2015.08.037

Kumar M, Yadav A (2015) Experimental investigation of solar powered water production from atmospheric air by using composite desiccant material "CaCl2/saw wood." Desalination 367:216–222. https://doi.org/10.1016/j.desal.2015.04.009

Kumar M, Yadav A (2017) Composite desiccant material '"CaCl2/Vermiculite/Saw wood"': a new material for fresh water production from atmospheric air. Appl Water Sci 7:2103–2111. https://doi.org/10.1007/s13201-016-0406-3

Kumar P, Anand B, Tsang YF, Kim KH, Khullar S, Wang B (2019) Regeneration, degradation, and toxicity effect of MOFs: opportunities and challenges. Environ Res 176:108488. https://doi.org/10.1016/j.envres.2019.05.019

LaPotin A, Kim H, Rao SR, Wang EN (2019) Adsorption-based atmospheric water harvesting: impact of material and component properties on system-level performance. Acc Chem Res 52(6):1588–1597. https://doi.org/10.1021/acs.accounts.9b00062
LaPotin A, Zhong Y, Zhang L, Zhao L, Leroy A, Kim H, Rao SR, Wang EN (2021) Dual-stage atmospheric water harvesting device for scalable solar-driven water production. Joule 5:166–182. https://doi.org/10.1016/j.joule.2020.09.008
Li R, Shi Y, Alsaedi M, Wu M, Shi L, Wang P (2018) Hybrid hydrogel with high water vapor harvesting capacity for deployable solar-driven atmospheric water generator. Environ Sci Technol 52:11367–11377. https://doi.org/10.1021/acs.est.8b02852
Liu X, Beysens D, Bourouina T (2022) Water harvesting from air: current passive approaches and outlook. ACS Materials Lett 4:1003–1024. https://doi.org/10.1021/acsmaterialslett.1c00850
Logan MW, Langevin S, Xia Z (2020) Reversible atmospheric water harvesting using metal-organic frameworks. Sci Rep 10:1492. https://doi.org/10.1038/s41598-020-58405-9
Loo SL, Vásquez L, Paul UC, Campagnolo L, Athanassiou A, Fragouli D (2020) Solar-driven freshwater generation from seawater and atmospheric moisture enabled by a hydrophilic photothermal foam. ACS Appl Mater Interfaces 12:10307−10316.https://doi.org/10.1021/acsami.9b20291
Lord J, Thomas A, Treat N, Forkin M, Bain R, Dulac P, Behroozi CH, Mamutov T, Fongheiser J, Kobilansky N, Washburn S, Truesdell C, Lee C, Schmaelzle PH (2021) Global potential for harvesting drinking water from air using solar energy. Nature 598:611–617. https://doi.org/10.1038/s41586-021-03900-w
Ma Q, Zheng X (2022) Preparation and characterization of thermo-responsive composite for adsorption-based dehumidification and water harvesting. Chem Eng J 429:132498. https://doi.org/10.1016/j.cej.2021.132498
Martens JA, Jacobs P (1994) Crystalline microporous phosphates: a family of versatile catalysts and adsorbents, in advanced zeolite science and application. Stud Sur Sci Catal 85:653–685. https://doi.org/10.1002/chin.199521300
McBain JW (2009) XCIX. The mechanism of the adsorption ("sorption") of hydrogen by carbon. Philos Mag Ser 6(18):916–935. https://doi.org/10.1080/14786441208636769
Moghadam PZ, Li A, Liu X-W, Bueno-Perez R, Wang S-D, Wiggin SB, Wood PA, Fairen-Jimenez D (2020) Targeted classification of metal–organic frameworks in the Cambridge structural database (CSD). Chem Sci 11:8373–8387. https://doi.org/10.1039/D0SC01297A
Moghaddam RH, Dadfarnia S, Shabani AMH, Tavakol M (2019) Synthesis of composite hydrogel of glutamic acid, gum tragacanth, and anionic polyacrylamide by electron beam irradiation for uranium (VI) removal from aqueous samples: Equilibrium, kinetics, and thermodynamic studies. Carbohydr Polym 206:352–361. https://doi.org/10.1016/j.carbpol.2018.10.030
Mohamed MH, William GE, Fatouh M (2017) Solar energy utilization in water production from humid air. Sol Energy 148:98–109. https://doi.org/10.1016/j.solener.2017.03.066
Mulchandani A, Westerhoff P (2020) Geospatial climatic factors influence water production of solar desiccant driven atmospheric water capture devices. Environ Sci Technol 54:8310–8322. https://doi.org/10.1021/acs.est.0c00534
Ng E-P, Mintova S (2008) Nanoporous materials with enhanced hydrophilicity and high water sorption capacity. Micropor Mesopor Mater 114:1–26. https://doi.org/10.1016/j.micromeso.2007.12.022
Nguyen GT, Hwang HS, Park I (2020) $MgSO_4$ composites in a ferroaluminophosphate for enhancement of volumetric heat storage capacity at a low charging temperature. Int J Energy Res 45:5177–5189. https://doi.org/10.1002/er.6132
Park H, Haechler I, Schnoering G, Ponte MD, Schutzius TM, Poulikakos D (2022) Enhanced atmospheric water harvesting with sunlight-activated sorption ratcheting. ACS Appl Mater Interfaces 14:2237–2245. https://doi.org/10.1021/acsami.1c18852
Permyakova A, Wang S, Courbon E, Nouar F, Heymans N, D'Ans P, Barrier N, Billemont P, Weireld GD, Steunou N, Frère M, Serre C (2017) Design of salt-metal organic frameworks composites for seasonal heat storage applications. J Mater Chem A 5:12889–12898. https://doi.org/10.1039/C7TA03069J

Polanyi M (1932) Section III.—theories of the adsorption of gases. A general survey and some additional remarks. Introductory paper to section III. Trans Faraday Soc 28:316–333. https://doi.org/10.1039/TF9322800316

Posern K, Linnow K, Niermann M, Kaps C, Steiger M (2015) Thermochemical investigation of the water uptake behavior of MgSO4 hydrates in host materials with different pore size. Thermochim Acta 611:1–9. https://doi.org/10.1016/j.tca.2015.04.031

Progress on level of water stress, Global Status and Acceleration Needs for SDG Indicator (2021) Food and agriculture organization of the United Nations, Rome. https://reliefweb.int/report/world/progress-level-water-stress-global-status-and-acceleration-needs-sdg-indicator-642. Accessed 27 Aug 2021

Rieth AJ, Yang S, Wang EN, Dinca M (2017) Record atmospheric fresh water capture and heat transfer with a material operating at the water uptake reversibility limit. ACS Cent Sci 3:668–672. https://doi.org/10.1021/acscentsci.7b00186

Ruthven DM (1984) Principles of adsorption and adsorption processes. Wiley, New York

Salehi AA, Ghannadi-Maragheh M, Torab-Mostaedi M, Torkaman R, Asadollahzadeh M (2020) A review on the water—energy nexus for drinking water production from humid air. Ren Sust Ener Rev 120:109627. https://doi.org/10.1016/j.rser.2019.109627

Schoenecker PM, Carson CG, Jasuja H, Flemming CJJ, Walton KS (2012) Effect of water adsorption on retention of structure and surface area of metal-organic frameworks. Ind Eng Chem Res 51(18):6513. https://doi.org/10.1021/ie202325p

Schüth F, Sing KSW, Weitkamp J (2002) Handbook of porous solids. Wiley-VCH, Weinheim, Germany

Seo Y-K, Yoon JW, Lee JS, Hwang YK, Jun C-H, Chang J-S, Wuttkee S, Bazin P, Vimont M, Daturi M, Bourrelly S, Liewellyn PL, Horcajada P, Serre C, Ferrey G (2012) Energy-efficient dehumidification over hierarchically porous metal-organic frameworks as advanced water adsorbents. Adv Mater 24:806–810. https://doi.org/10.1002/adma.201104084

Shiklomanov I (1993) Chapter 2, World fresh water resources. In: Gleick PH (ed) Water in crisis: a guide to the world's fresh water resources, 2nd edn. Oxford University Press, New York

Shkatulov A, Gordeeva LG, Girnik IS, Huinink H, Aristov YI (2020) Novel adsorption method for moisture and heat recuperation in ventilation: composites "LiCl/matrix" tailored for cold climate. Energy 201:117595. https://doi.org/10.1016/j.energy.2020.117595

Silva MP, Ribeiro AM, Silva CG, Nogueira IBR, Cho K-H, Lee U-H, Faria JL, Loureiro JL, Chang J-S, Rodrigues AE, Ferreira A (2021) MIL-160(Al) MOF's potential in adsorptive water harvesting. Adsorption 27:213–226. https://doi.org/10.1007/s10450-020-00286-5

Sleiti AK, Al-Khawaja H, Al-Khawaja H, Al-Ali M (2021) Harvesting water from air using adsorption material—prototype and experimental results. Sep Purif Technol 257:117921. https://doi.org/10.1016/j.seppur.2020.117921

Solovyeva M, Krivosheeva I, Gordeeva L, Aristov Y (2021a) MIL-160 as an adsorbent for atmospheric water harvesting. Energies 14:3586. https://doi.org/10.3390/en14123586

Solovyeva M, Krivosheeva I, Gordeeva L, Shkatulov A (2021b) Composites Salt/MIL-101(Cr) as adsorbents for water harvesting from the atmosphere. In: Abstracts of the 4th European conference on metal organic frameworks and porous polymers, Jagiellonian University, Kraków, 13–15 Sep 2021

Srivastava S, Yadav A (2018) Extraction of water particles from atmospheric air through a Scheffler reflector using different solid desiccants. Int J Ambient Energy. https://doi.org/10.1080/01430750.2018.1517667

Srivastava S, Yadav A (2020) Experimentally investigation of extraction of water vapours with Scheffler reflector through Nobel composite desiccant material. Int J Ambient Energy. https://doi.org/10.1080/01430750.2021.1873850

Talaat MA, Awad MM, Zeidan EB, Hamed AM (2018) Solar-powered portable apparatus for extracting water from air using desiccant solution. Renew Energy 119:662–674. https://doi.org/10.1016/j.renene.2017.12.050

(The) Right to Water, Fact Sheet No. 35. (2010) United Nations, OHCHR, UN-HABITAT, WHO

Thommes M, Kaneko K, Neimark AV, Olivier JP, Rodriguez-Reinoso F, Rouquerol J, Sing KSW (2015) Physisorption of gases, with special reference to the evaluation of surface area and pore size distribution (IUPAC Technical Report). Pure Appl Chem 87(9–10):1051–1069. https://doi.org/10.1515/pac-2014-1117

Tingting P, Kaijie Y, Yu H (2020) Recent progress of atmospheric water harvesting using metal-organic frameworks. Chem Res Chinese Univ 36(1):33–40. https://doi.org/10.1007/s40242-020-9093-6

Tranchemontagne DJ, Mendoza-Cortés JL, O'Keeffe M, Yaghi OM (2009) Secondary building units, nets and bonding in the chemistry of metal−organic frameworks. Chem Soc Rev 38:1257–1283. https://doi.org/10.1039/B817735J

Tu YD, Wang RZ, Zhang YN, Wang JY (2018) Progress and expectation of atmospheric water harvesting. Joule 2:1452–1475. https://doi.org/10.1016/j.joule.2018.07.015

Vidhi R (2018) A review of underground soil and night sky as passive heat sink: design configurations and models. Energies 11:2941. https://doi.org/10.3390/en11112941

Wada Y, de Graaf IEM, van Beek LPH (2016) High-resolution modeling of human and climate impacts on global water resources. J Adv Model Earth Syst 8:735–763. https://doi.org/10.1002/2015MS000618

Wang C, Liu X, Keser Demir N, Chen JP, Li K (2016) Applications of water stable metal–organic frameworks. Chem Soc Rev 45:5107–5134. https://doi.org/10.1039/C6CS00362A

Wang JY, Liu JY, Wang RZ, Wang LW (2017a) Experimental investigation on two solar-driven sorption based devices to extract fresh water from atmosphere. Appl Therm Eng 127:1608–1616. https://doi.org/10.1016/j.applthermaleng.2017.09.063

Wang JY, Wang RZ, Wang LW, Liu JY (2017b) A high efficient semi-open system for fresh water production from atmosphere. Energy 138:542–551. https://doi.org/10.1016/j.energy.2017.07.106

Wang JY, Wang RZ, Tu YD, Wang LW (2018) Universal scalable sorption-based atmosphere water harvesting. Energy 165:387–395. https://doi.org/10.1016/j.energy.2018.09.106

Wang X, Li X, Liu G, Li J, Hu X, Xu N, Zhao W, Zhu B, Zhu J (2019) An interfacial solar heating assisted liquid sorbent atmospheric water generator. Angew Chem Int Ed 58:12054–12058. https://doi.org/10.1002/ange.201905229

Wang W, Xie S, Pan Q, Dai Y, Wang R, Ge T (2021a) Air-cooled adsorption-based device for harvesting water from island air. Renew Sustain Energy Rev 141:110802. https://doi.org/10.1016/j.rser.2021.110802

Wang M, Sun T, Wan D, Dai M, Ling S, Wang J, Liu Y, Fang Y, Xu S, Yeo J, Yu H, Liu S, Wang Q, Li J, Yang Y, Fan Z, Chen W (2021b) Solar-powered nanostructured biopolymer hygroscopic aerogels for atmospheric water harvesting. Nano Energy 80:105569. https://doi.org/10.1016/j.nanoen.2020.105569

Wang Y, Gao S, Zhong H, Zhang B, Cui M, Jiang M, Wang S, Wang Z (2022) Heterogeneous wettability and radiative cooling for efficient deliquescent sorbents-based atmospheric water harvesting. Cell Reports Phys Sci 3:100879. https://doi.org/10.1016/j.xcrp.2022.100879

Wasti TZ, Sultan M, Aleem M, Sajjad U, Farooq M, Raza HMU, Khan MU, Noor S (2022) An overview of solid and liquid materials for adsorption-based atmospheric water harvesting. Adv Mech Eng 14(3):1–27. https://doi.org/10.1177/16878132221082768

Wei DD, Yang J, Zhu L, Chen F, Tang ZQ, Qin G, Chen Q (2018) Semicrystalline hydrophobically associated hydrogels with integrated high performances. ACS Appl Mater Interfaces 10:2946–2956. https://doi.org/10.1021/acsami.7b15843

William GE, Mohamed MH, Fatouh M (2013) Simulation of a water recovery from atmospheric air system by using solar energy. ICFD11, Alex, Egypt

William GE, Mohamed MH, Fatouh M (2015) Desiccant system for water production from humid air using solar energy. Energy 90:1707–1720. https://doi.org/10.1016/j.energy.2015.06.125

Wilson ST, Lok BM, Messina CA, Cannan TR, Flanigen EM (1982) Aluminophosphate molecular sieves: a new class of microporous crystalline inorganic solids. J Am Chem Soc 104:1146–1147. https://doi.org/10.1021/ja00368a062

Wu J, Zhou YM, Meng Y, Zhang JX, Liu QB, Cao QM, Yu YQ (2015) Synthesis and properties of sodium alginate/poly(acrylic acid) double-network superabsorbent. E-Polym 15:271–278. https://doi.org/10.1515/epoly-2015-0060
Xu W, Yaghi OM (2020) Metal-organic frameworks for water harvesting from air, anywhere, anytime. ACS Cent Sci 6:1348–1354. https://doi.org/10.1021/acscentsci.0c00678
Xu JX, Li TX, Chao JW, Yan TS, Wang RZ (2019) High energy-density multi-form thermochemical energy storage based on multi-step sorption process. Energy 185:1131–1142. https://doi.org/10.1016/j.energy.2019.07.076
Xu J, Li T, Chao J, Wu S, Yan T, Li W, Cao B, Wang R (2020) Efficient Solar-driven water harvesting from arid air with metalorganic frameworks modified by hygroscopic salt. Angew Chem Int Ed 59(13):5202–5210. https://doi.org/10.1002/anie.201915170
Xu XW, Jerca VV, Hoogenboom R (2021) Bioinspired double network hydrogels: from covalent double network hydrogels via hybrid double network hydrogels to physical double network hydrogels. Mater Horiz 8:1173–1188. https://doi.org/10.1039/D0MH01514H
Yaghi OM, O'Keeffe M, Ockwig NW, Chae HK, Eddaoudi M, Kim J (2003) Reticular synthesis and the design of new materials. Nature 423:705–714. https://doi.org/10.1038/nature01650
Yaghi OM, Kalmutzki MJ, Diercks CS (2019) Introduction to reticular chemistry: metal-organic frameworks and covalent organic frameworks. Wiley-VCH Verlag GmbH & Co, KGaA
Yang RT (2003) Adsorbents: fundamentals and applications. Willey, New Jersey
Yang K, Pan T, Lei Q, Dong X, Cheng Q, Han Y (2021) A roadmap to sorption-based atmospheric water harvesting: from molecular sorption mechanism to sorbent design and system optimization. Environ Sci Technol 55(10):6542–6560. https://doi.org/10.1021/acs.est.1c00257
Yu Q, Zhao H, Sun S, Zhao H, Li G, Li M, Wang Y (2019) Characterization of MgCl2/AC composite adsorbents and its water vapor adsorption for solar drying system application. Renew Energy 138:1087–1095. https://doi.org/10.1016/j.renene.2019.02.024
Zhang Y, Wang R, Li T, Zhao Y (2016) Thermochemical characterizations of novel vermiculite-LiCl composite sorbents for low-temperature heat storage. Energies 9:854. https://doi.org/10.3390/en9100854
Zhang Z, Fu H, Li Z, Huang J, Xu Z, Lai Y, Qian X, Zhang S (2022) Hydrogel materials for sustainable water resources harvesting & treatment: synthesis, mechanism and applications. Chem Eng J 439:135756. https://doi.org/10.1016/j.cej.2022.135756
Zhao F, Zhou X, Liu Y, Shi Y, Dai Y, Yu G (2019) Super moisture-absorbent gels for all-weather atmospheric water harvesting. Adv Mater 31:1806446. https://doi.org/10.1002/adma.201806446
Zhao H, Lei M, Liu T, Huang T, Zhang M (2020) Synthesis of composite material HKUST-1/LiCl with high water uptake for water extraction from atmospheric air. Inorganica Chim Acta 511:119842. https://doi.org/10.1016/j.ica.2020.119842
Zhao H, Huang T, Liu T, Lei M, Zhang M (2021) Synthesis of $MgCl_2$/vermiculite and its water vapor adsorption-desorption performance. Int J Energy Res 2021:1–15. https://doi.org/10.1002/er.7188
Zhou X, Lu H, Zhao F, Yu G (2020) Atmospheric water harvesting: a review of material and structural designs. ACS Mater Lett 2:671–684. https://doi.org/10.1021/acsmaterialslett.0c00130

Chapter 4
Metal-Oxide Frameworks for Atmospheric Water Harvesting

Shatakshi Srivastava, Tanushri Chatterji, Namrata Khanna, Suruchi Singh, Kwena D. Modibane, Orebotse Joseph Botlhoko, Edwin Makhado, and Sadanand Pandey

4.1 Introduction

The rapidly rising population, climate change, and pollution in the twenty-first century have all put a strain on fresh water resources across the world. As a result, more scientists have drawn attention to the need for innovative renewable water harvesting systems than ever before (Boretti and Rosa 2019; Unicef 2020). Though

S. Srivastava
Apeejay Stya University, Sohna-Palwal Road, Gurugram 122103, India

T. Chatterji (✉)
School of Bioscience, Institute of Management Studies (IMS) Ghaziabad (University Courses Campus), Ghaziabad 201015, India
e-mail: tanushri@gmail.com

N. Khanna
Department of Biochemistry, M A Rangoonwala College of Dental Sciences and Research Centre, 2390-B, K.B. Hidayatullah Road, Azam Campus, Camp, Pune 411001, India

S. Singh
Life Sciences, Tata Consultancy Services Limited, Hiranandani Estate, Thane (W), Mumbai 400607, India

K. D. Modibane · E. Makhado (✉)
Department of Chemistry, School of Physical and Mineral Sciences, University of Limpopo, Sovenga, Polokwane 0727, South Africa
e-mail: edwin.makhado@ul.ac.za

O. J. Botlhoko
Centre for Nanostructures and Advanced Materials, DSI-CSIR Nanotechnology Innovation Centre, Council for Scientific and Industrial Research, Pretoria 0001, South Africa

S. Pandey (✉)
Department of Chemistry, College of Natural Science, Yeungnam University, 280 Daehak-Ro, Gyeongsan, Gyeongbuk 38541, Republic of Korea
e-mail: spandey@ynu.ac.kr; sadanand.au@gmail.com

E. Fosso-Kankeu et al. (eds.), *Atmospheric Water Harvesting Development and Challenges*, Water Science and Technology Library 122,
https://doi.org/10.1007/978-3-031-21746-3_4

seawater constitutes 97.5% of the total water, fresh water is only about 2.5%. Out of this 2.5%, only 0.4% is available in ready to use form and the rest is present in the form of fog, dew, atmospheric vapours, rivers, lakes etc. (Bilal et al. 2022), which limits the availability of water to many locations globally. The total moisture content in the Earth's atmosphere is nearly 13,000 km^3, greatly exceeding global demands (Khalil et al. 2016; Domen et al. 2014). Because the atmosphere is abundant, atmospheric water harvesting has occurred since ancient times. Atmospheric Water Harvesting (AWH) is being viewed as an alternative to desalination technology in remote arid regions. AWH has been shown to be a feasible alternative for providing fresh drinking water. Harvesting water from ambient air has the potential to be largely powered by renewable energy sources. The passive mode of water collection involves collecting fog and/or dew water from the atmosphere without any external energy source. In contrast to fog, which is produced by the evaporation of water or the sublimation of ice, dew is formed when water condenses on a cool surface below the dew point temperature (Khalil et al. 2016; Domen et al. 2014).

The fog water collection includes the traditional method of the fog collector and the modern method of bio-mimetic inspired fog water. Dew water harvesting can be accomplished using a radiative cooling condenser (less efficient), a solar regenerated desiccant, or active cooling condensation technology, which traditionally uses vapour compression air conditioning systems. Though fog collection is an interesting and well-rewarding technique, it suffers from a setback that its large water collection duration is limited to the monsoon season only. i.e., two-to-three months a year. Other limitations include lack of rational designing of the mesh type components of the fog harvester, causing weakening or broadening of the mesh due to prevailing winds. The Warka water and cloud harvesters with steel mesh are positive modifications of technology that assures minimum wear and tear and efficiently collects and condenses the fog into water droplets. In Chile, two fog collectors: a standard fog collector and one fog collector with a local design were installed by a local fisherman's association at 600 m height at the "Falda Verde" site from November 1998 to November 2000. In the local fog design, a 1.5 m^2 fishing type mesh was mounted on wooden poles, which were planted 1 m above the ground. A second unit was installed two metres apart, having 1 m^2 mesh, 35% coverage, and 2 m above the ground. These fog collectors produced an average of 1.46 l/m^2 per day (Larrain et al. 2001, 2002). In 2005, four large fog collectors in the Tojquia, Western Highlands of Guatemala were constructed, and the number was increased to 35 in forthcoming years. This project is one of the most successful fog harvesting projects so far, with a water production of 7000 L per day and providing potable water to the village (Schemenauer et al. 2016). Many other fog projects across the globe are installed and are giving satisfactory results, but for the afore-mentioned reasons, newer modifications are welcome (Rojas et al. 2014; Calderón et al. 2010; Jaen 2002).

The modern development in fog water harvesting methods involves bio-mimicry-inspired fog water harvesting. This method takes its inspiration from the natural ability of various animals to collect and condense water even under harsh conditions, e.g., Stenocana gracilipes found in the Namib deserts survives successfully with only

12 mm of annual rainfall (Nørgaard and Dacke 2010). Its water-harvesting adaptations include a random assortment of small, smooth hydrophilic bumps and grooves that appear on its back occasionally. Through this, water can condense and drip right into the animal's mouth. Different adaptations for hydrophobic and hydrophilic areas in the hairs, grooves, and barbs of insects, beetles, frogs, lizards, and spiders are responsible for efficient fog water harvest and dew water harvesting. Similar to this, microgrooves and microscale tilted cones exhibit asymmetric anisotropic directional mist collecting behavior, nanoscale coated surfaces, and microfluidic technology to control flow rate in plants (e.g., *Trianthema hereroensis, Stipagrostis sabulicola,* and *Opuntia microdasys.*) resulting in the efficient capture of atmospheric water (Tu et al. 2018; Garrod et al. 2007; Brown and Bhushan 2016; Zheng et al. 2010). These adaptations were studied by numerous researchers and applied to different materials for preparing atmospheric water generators (Inbar et al. 2020; Suvindran et al. 2018; Jalali et al. 2021).

The occurrence of global fog is determined by geographical and meteorological factors or conditions, and the significant release of heat during water condensation is one of the major limitations of fog water harvesting. Therefore, dew water harvesting seemed to be a more befitting supplement to fog harvesting. The techniques for gathering it may be broadly classified into three groups: (a) passive or radiative cooling condensation (PCC); (b) solar-regenerated desiccant; and (c) active cooling condensation technology from air (Khalil et al. 2016; Beysens et al. 2007; Alnaser and Barakat 2000). The passive cooling condenser or radiative cooling harvest system is based on the principle that water deposits on plant surfaces which have been cooled down to the dew-point temperature of the surrounding air by losing heat in the form of radiation (Baier 1966). The cooling of the surface and water condensation in passive cooling are affected by factors like infrared wavelength emitting properties of the surface, reflectivity of the condensing surface, wind effect on the condenser, heat inertia of the condensing surface, and hydrophilic properties of the surface (Liu et al. 2022; Sharan 2011; Carvajal et al. 2018; Nioras et al. 2021). The passive systems are driven only by solar energy, but lower yields are a limiting factor of this technique.

Another improvised version of passive dew harvesting utilizes solar regenerated desiccant materials like calcium chloride, sawdust, recycled newspaper, zeolites, and silica gel (Gado et al. 2022; Wang et al. 2016). These desiccant materials exhibit both the mechanisms of absorption and thereby efficiently increase their water retention ability. Desiccant beds are prepared for AWH, which involves the water being absorbed/adsorbed from humid air till saturation at night and desorption taking place during the day. The water evaporated during the day is condensed and finally collected in collection tanks (Wang et al. 2016). Porous metal–organic framework (Suh et al. 2019; Wu et al. 2021; Hanikel et al. 2021), corrugated cloth surface (Gad et al. 2001), glass pyramid collector (Kabeel 2007), solar glass desiccant box type system (Kumar and Yadav 2015) and collection tank structural designs play an important role in long-term water intake and recyclability (Zhou et al. 2020).

The advancement in technology led to active water harvesting methods, which need an additional source of energy input to harvest more water than passive methods (Gido et al. 2016; Sharan et al. 2017). Active water harvesting systems need external energy sources for membrane-assisted harvesting, vapour-compression, and Peltier or thermoelectric cooling. The active dew condensers work on the principle of a dehumidifier in which cooling of moist air by a refrigerant-cooled coil leads to water condensation. Atmospheric water generators utilize photovoltaic cells or sorption-based thermal techniques for entrapping solar energy as the external source for harvesting moisture from air. Different modifications to heating, ventilation, and air-conditioning systems are being proposed from time to time to increase the generation of water from the atmosphere. These modifications are based on the fact that the RH and temperature of air directly correlate with the amount of energy consumed for atmospheric water harvesters. The weather conditions of each area require a particular type of atmospheric water generator for effective harvesting of water from the atmosphere. For example, if the daytime humidity of an area is as low as 10%, then only 3 L of water can be processed from 1 million litres of air, while the night time humidity is about 40% and the temperature falls below – 20 °C, which prevents water harvesting by the refrigeration-based method. Thermodynamically speaking, the more energy efficient the process is for AWH, the more cost effective, safe, and stable the AWH technology will be. Different studies on various AWH technologies have indicated that adsorption-based harvesting is an energy-efficient, cost-effective, and environmentally friendly technique. Primarily, we will discuss the adsorption material, designs, and thermodynamics used for adsorption based AWH technology.

4.2 Atmospheric Water Harvesting Based on Adsorption

Chillers, which use adsorption-based technology, extract water from the air when the RH is low. Adsorption-based AWH systems differ from conventional AWH systems in that they use desiccant materials to collect water vapor from the air and exhibit improved thermal efficiency. These raise the pressure of the system for producing water by using reasonably priced, clean solar energy or waste energy. Three processes make up the water harvesting cycle: adsorption at night, desorption during the day, and condensation. When the RH is high and the temperature is low, ambient water vapor is absorbed by the adsorption bed. This is exposed to sunlight during the day time for the water desorption process to occur, and then the high vapour content generated during desorption is condensed into water droplets. This makes AWH systems work efficiently even in low RH areas.

Before going further, let us first understand the parameters of water vapour, viz., absolute humidity (ω), dew point temperature (Td) and RH (Φ), for its harvesting from the atmosphere. RH (Φ) is the ratio of the partial pressure of water vapor (Pw) to the saturation pressure (Ps), the vapour pressure that has attained saturation. It can be expressed as (Petersen et al. 2016):

$$\Phi = \frac{P_\omega}{P_s(T)}$$

Regardless of the air's temperature, absolute humidity is a measurement of the actual amount of water vapor (moisture) present. The ratio of absolute humidity to the amount of water vapor in the atmosphere is direct. The correlation between the three parameters is given by the following equation:

$$\Phi = \frac{\omega P}{(0.622 + \omega) P_s(T)}$$

where P stands for total pressure and T stands for temperature

The dew point is the temperature needed to cool air to 100% RH (RH) while maintaining constant pressure. The ability of air to contain water at this point is zero. It can be calculated by taking the temperature T at constant pressure and humidity and keeping $\Phi = 1$. Generally, the rate of energy consumption increases with the decrease in RH in atmospheric water harvesters, but in adsorption based AWH, the heat and body temperature are constant, making it a better choice. Its efficiency can also be manipulated as needed by selecting the appropriate sorption characteristics such as isotherm shape and step position, thermal conductivity, binding energy, and saturation capacity (Gado et al. 2022; Gad et al. 2001; Gido et al. 2016; Mabokela et al. 2022; Kim et al. 2018; LaPotin et al. 2019). The performance of AWG systems is based on various parameters like specific water production per day per unit collector area, specific energy consumption (SEC) per unit mass of water collected, relative pressure (RP), recovery ratio (RR) of feed air, heat of regenerated air (W_{heat}), adsorption capacity per unit mass of adsorbent (Δx), the design of the sorbent bed structure, and the system's water harvesting capacity ($M_{water,}$ duration of adsorption and desorption) (Bilal et al. 2022; Tu et al. 2018; LaPotin et al. 2019; Kim et al. 2020; Wang et al. 2018). The pores of the sorbent play an important role in water condensation. The pore filling must occur at a low RH as water is captured from the atmosphere where moisture content is too low. It should also display a steep uptake behavior for better adsorption capacity. Microporous zeolites fulfil all the above criteria and may serve as good adsorbents. However, their recyclability requires a high amount of energy. Recent research has shown the use of metal–organic frameworks (MOFs) for moisture capture as a more energy efficient method. The values of SEC and RR for direct cooling water collection may be defined as follows:

$$SEC = \frac{Q_{cond}}{mH_2O} \approx C_p\left(\frac{\varepsilon_T}{\varepsilon_d}\right)\left(\frac{T_i - T_{cond}}{d_i - d_{cond}}\right) + h_{fg},$$

$$RR = \varepsilon_d\left(1 - \frac{d_{cond}}{d_i}\right),$$

where T_i and d_i represent the incoming air's temperature and RH, respectively. T_{cond} is the condensation temperature. mH_2O is the measure of water production per kilogram of dry mass air (kg/kg). ε_T stands for the condenser's efficiency in exchanging heat, whereas ε_d denotes its efficiency in exchanging mass. The entire cooling load of wet air is Q_{cond}(= sensible heating load + latent heat load) and h_{fg} is the enthalpy of condensation (Tu et al. 2018). Also, each adsorbent undergoes two processes; adsorption (T_{ad}, P_{ad}) and desorption (T_{de}, P_{de}). The following equation gives the water collecting capacity per unit mass Δx:

$$\Delta x = x_{ad}(T_{ad}, RP_{ad}) - x_{de}(T_{de}, RP_{de})$$

$$RP = \frac{P_v}{P_{sat}(T_{sorbent})} = RH \times \frac{P_{sat}(T_{air})}{P_{sat}(T_{sorbent})}$$

$$M_{water} = p_{air,sorber} Q_{air,sorber} t_{sorber} (d_{desorber,o} - d_{cond})$$

$$RR = \frac{M_{water}}{p_{air,sorber} Q_{air,sorber} t_{sorber} d_{sorber,i}} \approx \frac{d_{desorber,o} - d_{cond}}{d_{sorber,i}}$$

$$W_{heat} = C_p P_{air,desorber} Q_{air,desorber} t_{desorber} (T_{desorber,i} - T_{ambient})$$

where P_v = water vapor's partial pressure, P_{sat} is saturated vapor pressure, andRH is theRH of the bulk air. Saturated humidity ratio at condensing temperature T_{cond} = d_{cond}. $d_{desorber,o}$ = moisture content of the desorber outlet air and dsorber, i = sorber inlet moisture content (Bilal et al. 2022).

Due to the low rate of reaction, the desiccant temperature Tad is almost equivalent to the bulk air temperature T_{air} for the adsorption process, $RP_{ad} \approx RH_{ad}$. A low RP_{de} and a high T_{de} are beneficial for the desorption process because they help remove moisture from the desiccant. The greater sorber heat transfer efficiency will be feasible with a lower value of RP_{de} and a higher T_{de} if the regeneration air itself heats the desiccant (Wang et al. 2018). This implies that advancements in desiccant material, properties such as crystal size, thermal conductivity, vapor transport to a condenser, and design can have a significant impact on water production. Nowadays, adsorption-based AWGs frequently employ nanoporous materials such as metal organic frameworks, organic and inorganic hygroscopic materials including LiCl, $CaCl_2$, composite materials, and zeolites (Sharan 2011; Wang et al. 2016; Gad et al. 2001; Zhou et al. 2020).

The adsorption based AWH is based on adsorption-desorption process where an adsorbate and absorbent is required. Water is the adsorbate, and zeolites, silica, and metal organic frameworks (MOFs) are common examples of highly heterogeneous and porous adsorbent materials. Instead of diffusing, the process of adsorption involves the attachment of gas or liquid molecules (adsorbate) to the surface of an adsorbent. Desorption refers to the separation of these molecules from the adsorbent surface. Adsorption is of two types: physisorption and chemisorption. Whereas chemisorption involves formation of chemical covalent bonds at binding sites with high enthalpy change, the physisorption relies on weak interactions like Vander Waals forces for adsorbate- adsorbent interaction with lower enthalpy change. Depending upon the interactions and change in enthalpy involved, three basic types of adsorption mechanisms are observed so far for MOFs and water interaction. These are: (i) adsorption on open metal sites by chemisorption (modifies the metal ion coordination sphere), (ii) irreversible, discontinuous capillary condensation and (iii) reversible continuous pore filing mechanism (Canivet et al. 2014a). Apart from Langmuir isotherm (monolayer surface), Freundlich isotherm (heterogenous surface), Brunauer Emmett Teller (BET) isotherms (multilayer surface), LDF (linear driving force model) has also been suggested to describe the adsorption kinetics, which incorporates pseudo-first order and pseudo-second order kinetic equations.

In the middle of the 1980s, pseudo-second order kinetics was developed, while pseudo-first order kinetics was initially postulated by Lagergren in the late nineteenth century. Additional research revealed that the best connection with experimental data was obtained by pseudo-second order kinetics. Following that, studies on MIL-101, MIL-101(Cr) doped with alkali metals (Li^+, Na^+), aluminium fumarate, and CAU-10 were carried out to better understand the behavior of MOFs in water and their potential as an efficient adsorbent. According to its bigger pores and greater surface area, MIL-101 had the maximum water absorption of these three in the saturated pressure range. In addition, CAU-10 and aluminium fumarate have significantly lower hydrophobic lengths and less water absorption than MIL-101. Because the pore size distribution graph of the MOF materials revealed at least two distinct expandable distribution peaks, the MIL-101 samples have superior surface properties to CAU-10 and Aluminium Fumarate. The kinetics data of the modified MIL-101 demonstrated a superior capacity to attain a steady state more quickly. Therefore, MOF adsorbents are good candidates for water adsorption applications due to their larger uptake difference and quicker kinetic performance (Teo and Chakraborty 2017).

Metal–organic frameworks (MOFs) are a great contender for AWH because of their high adsorption capacity and potential for target-specific design. As porous materials made of charged organic ligands and metal ions, MOFs are often referred to as porous coordination polymers (Zhang et al. 2020). The inorganic components known as metal ions, or secondary building units (SBUs), are what define the shape of the MOF matrix based on its coordination number (C.N.). If the SBU has the coordination number of six, then it will exhibit the octahedron geometry or triagonal bipyramidal, for C.N. = 4, square paddle wheel and with C.N. = 3, the geometry will be triangular (Frameworks 2016). The organic part of MOFs acting as linker consists of anions like carboxylate, phosphonates, sulfonates and also heterocyclic compounds. However, the linker also modifies the geometry of MOF to some extent as it may attach to the metal ion through more than one labile sites depending upon its nature. The porous nature of MOFs is due to the large organic linker molecules that provides large guest molecule storage space and numerous adsorption sites within MOFs. The variation in size and shape of the linker molecule affects the pore diameter, its affinity to guest molecules and pore volume. The large linker molecules and high coordination metal atoms will result in the formation of large spaces in the MOF framework, resulting into interpenetrating structures. These interpenetrating structures should be avoided as these will affect the adsorption efficiency of the MOF. Optimum pore size and large porosity plays an important role in defining the function of MOF as a catalyst, sensor, drug-delivery agent, atmospheric water harvestor etc.

Recent research developed six new zirconium MOFs and compared them to zirconium MOF-801, UiO-66, MOF-804, DUT-67, and PIZOF-2. The new zirconium MOFs were MOF-802, MOF-805, MOF-806, MOF-808, MOF-812, and MOF-841.

Water from the atmosphere might be captured and released using MOF-841. The limited water absorption at reduced pressure indicates the poor affinity of water for the MOF surface. More water vapor pressure is required to promote pore filling due to the organic linker's associated hydrophobicity. This was proven by the behavior of PIZOF-2, a material with large pores that exhibits resistance to water absorption up to a predetermined relative pressure of $P/P0 = 0.7$ (where P0 is the pressure at which water reaches saturation). Examples include the type I isotherm of MOF-802 (Fig. 4.1a), the water isotherms for the principal stages of group 2 MOFs with a hysteresis loop (Fig. 4.1b), and the type I isotherm profile of group 3 MOFs for MOF-804 (Fig. 4.1c) (Furukawa et al. 2014). After the absorption starts, at $P/P0 = 0.9$, it reaches a complete water storage capacity of 850 cm^3/g. This high capacity for absorption at high relative pressure suggests that total water intake and porosity have a substantial connection (Furukawa et al. 2014). By altering the pore size and activation methods, the functionality of the MOF can be tailored (Llewellyn et al. 2008). On the basis of pore size, these could be defined as micropores (widths < 2 nm), mesopores (widths between 2 and 50 nm), macropores (> 50 nm). The adsorption isotherms of nitrogen and argon gases are generally calculated for studying the porosity of material at their boiling points. The pore size result in structural modifications in MOFs which consequently leads to alterations in adsorption isotherms. Therefore, water like adsorbate molecules display different types of isotherms and shape of each isotherm depends on physical and chemical properties of MOF adsorbent. Each physisorption isotherm of MOF with gases provide information about pore size, the matrix of pores and surface area. The BET (Brunaeur-Emmett-Teller) method is commonly applied to evaluate the surface area of MOFs and is a fingerprint of each MOF. If the isotherm has a horizontal plateau then the external surface area of MOF is very less and it does not contain large pores (Zhang 2016). The plateau in adsorption isotherm is formed when MOFs are saturated with adsorbed molecules e.g., Type I isotherms for microporous MOF and Type 1 V isotherm for mesoporous MOF (Fig. 4.1d).

4.3 Development of MOFs

The term MOF was proposed by Omar Yaghi in the year 1995 as porous structures in which the organic linkers or bridging ligands are attached to each other by dative bond. Its framework has permanent porosity due to large organic linkers and large surface area due to numerous pores. The porous network of these MOFs may form 1D, 2D or 3D crystalline structures. In 1989, Hoskins and Robson prepared first MOF $CuC(C6H4 \cdot CN)4_n^{n+}$ with copper as central metalion (Hoskins and Robson 1989). The porosity, crystalline nature, high stability, tunable metrices, extra-large surface area and organic functionality attracted newer synthesis approaches, like reticular synthesis, of these MOFs (Yaghi et al. 2003). The reticular synthesis provides the flexibility of manipulating the organic and inorganic ligands, geometries, alterations in structures yet keeping the topology constant. Consequently, over one lac structures

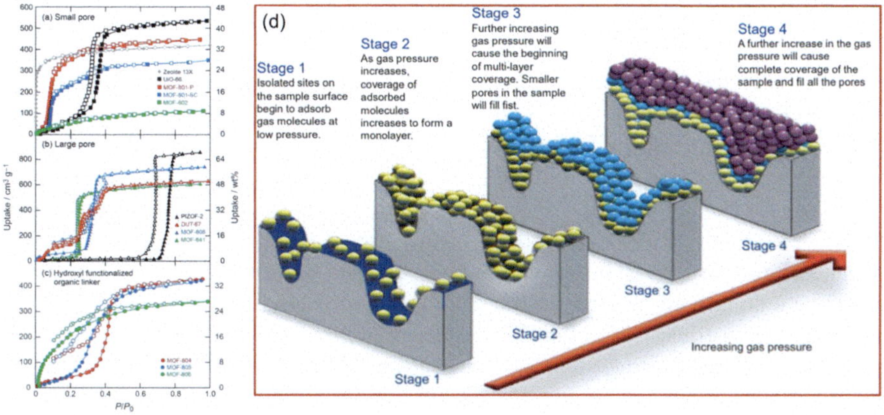

Fig. 4.1 Water isotherms of zirconium MOFs with small (**a**), large pores (**b**), hydroxyl functionalized linkers at 25 °C (**c**) (Reprinted with permission from Furukawa et al. (2014) Copyright 2014 American Chemical Society), and the diagram shows the adsorption stages of gases on an adsorbent (**d**) (Zhang 2016)

were reported in Cambridge Structural Database (CSD) in past few decades (Li et al. 2020).

Kitagawa et al. reported that metal–organic polymers canreversibly adsorb gas phase molecules in the 3D Framework with small molecule channeling cavities (Kondo et al. 1997). Rigid frameworks that preserve their structural integrity and porosity throughout anion-exchange, guest sorption from solution, and in the absence of guest molecules were developed by Li et al. (1998). Since then numerous researches have shown the role of metal organic frameworks in molecule separations (Zhou et al. 2020), drug discovery (Lawson et al. 2021; Jabalia et al. 2021), catalysis, biomolecule encapsulation (Xing et al. 2020) and more recently adsorption (Bilal et al. 2022; Canivet et al. 2014a; Furukawa et al. 2014; Ejeian and Wang 2021). Gas adsorption by MOFs proved to be an effective step towards creating permanently porous MOFs. The development of reticular synthesis in MOFs with the help of transition metals revolutionized the field by creating requirement specific MOFs. Zirconium containing metal frameworks in single crystal and crystalline forms are recently being used as adsorbents for water harvesting. For instance, according to Kim et al. (2017), MOF-801 could extract 2.8 L of water per kg of MOF per day at a RH of 20% of the air under ambient circumstances. Importantly it should be noted that MOF-801 follow repaid water harvesting strategy, reaching immature saturation point or early saturation point at about 80 min. This it is in relation to an inflection point at $P/P_0 \leq 0.1$ and plateau which is reached at about $P/P_0 = 0.2$. While MOF-841 takes time to capture water but get saturated quite late with instant water capturing mechanism (plateau is reached at $P/P_0 = 0.3$, and good temperature response is observed) (Kim et al. 2017; Hu et al. 2022).

The relevance of collecting water from atmospheric air was highlighted by the researchers since it was revealed that there is around 50% RH (RH) that may be gathered using metal organic frameworks (MOFs) (Suh et al. 2019; Kim et al. 2018, 2017; Xu and Yaghi 2020; Akiyama et al. 2012; Ko et al. 2015; Jeremias et al. 2012). They demonstrated that the temperature necessary for water molecules to saturate the air is extremely low (Fig. 4.3a). MOFs appear to be on the cutting edge of structured materials for water adsorption from air. The MOF was demonstrated to be capable of adsorbing water at 25 °C, indicating a reactive mechanism for water interaction with MOF pores. As a result, MOFs are regarded as the next-generation in water harvesting technology due to their large surface area, low density, customizable pore size, and a wide range of RH (Xu and Yaghi 2020). The metal core and organic ligands used have a significant impact on the structural characteristics of the MOF. MOFs are frequently porous, which lends themselves to their wide range of applications in catalysis, energy storage, and gas adsorption systems. MOF-303, for example, was employed to produce water in many places with varying weather conditions (Fig. 4.2b) (Xu and Yaghi 2020). Due to their unique wet and dry seasons, cities with a tropical savanna environment (Chennai and Dhaka) were shown to have high water production rates (Xu and Yaghi 2020). It was discovered that in the summer, Delhi, a city with a humid subtropical environment, water was produced in great quantities. On the other hand, water harvesting devices in Cities with Mediterranean climates include Los Angeles, Cape Town, Perth, Rome, and Granadarevealed that water adsorption during the summer was equivalent to that during the winter (Xu and Yaghi 2020).

In another study, Xu and Yaghi (2020) reported the usage of Zr-based MOFs for water adsorption as another example. Figure 4.4 depicts their observations on MOF water uptake. In comparison to MOF-303 (550 cm^3/g) and MOF-801 (350 cm^3/g), MOF-841 had the highest water adsorption–desorption capacity of roughly 600 cm^3/g (Fig. 4.3a). They also discovered good adsorption–desorption stability that lasted at least 80 cycles. At ambient temperature, these Zr-based MOFs adsorb water with a step-shaped isotherm, indicating water molecule binding into their pores (Xu and Yaghi 2020). They have produced a proof-of-concept device to demonstrate MOF's potential to adsorb water molecules from low-RH air outdoors. Their device was a simple glass "jar" with a cap filled with MOF that worked at night by exposing

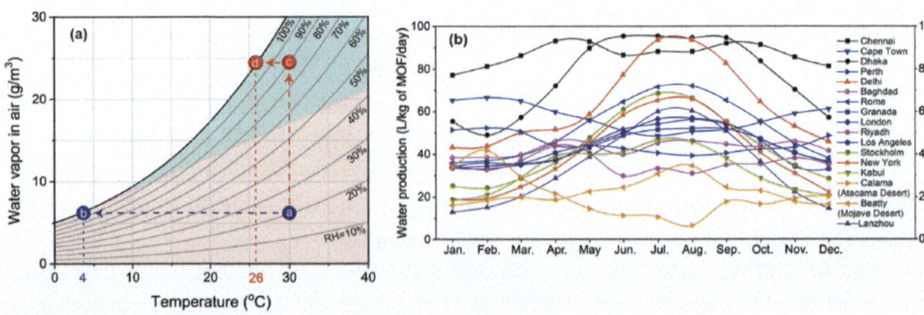

Fig. 4.2 **a** Graphic representation of low RH; and **b** water productivity of MOF-303 (Xu and Yaghi 2020)

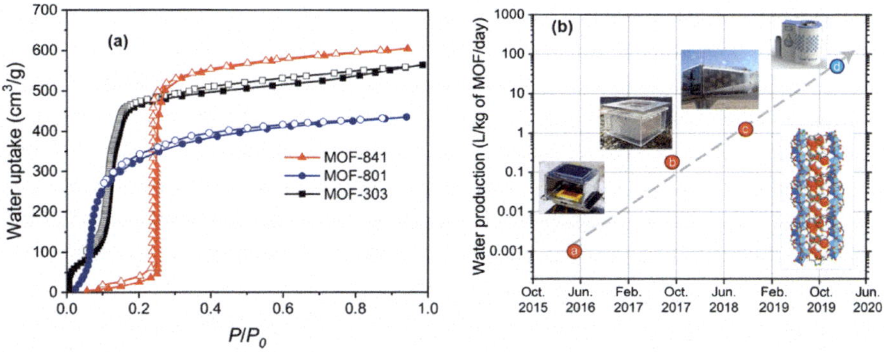

Fig. 4.3 **a** Water uptake properties of MOF-841, MOF-801, and MOF-303 materials for water adsorption from air; and **b** proof-of-concept device for water adsorption from air by MOF (Xu and Yaghi 2020)

the coated area to air and allowing it to absorb water (Fig. 4.3b). After adsorption, the device was shut and exposed to sunlight, at which point water was released from the MOF and condensed. The other experiment used only 2 g of MOF-801, which can absorb 10–100 L of water per 1 kg of MOF, and was carried out in the Arizona desert at % RH and 25 °C.

Suh and co-workers (2019) investigated the use of MOF (Ni-IRMOF74-III) incorporated with cis/trans transition of azopyridine molecules via photochemical process for water adsorption, Figure. At 25 °C and 2 kPa, they measured the water absorption of 0.33 kg/kg H_2O. (Fig. 4.4a). Trans and cis H_2O adsorptions were seen to converge on one another at greater pressures, indicating that these configurations absorb roughly the same quantity of water (Fig. 4.4b).

Different MIL-101 Cr materials with various substituents (–H, $-NO_2$, $-NH_2$, –SO_3H) on the organic linker were produced by Akiyama et al. (2012). According to the authors, the water adsorption capabilities ranged from 0.8 to 1.2 g of water

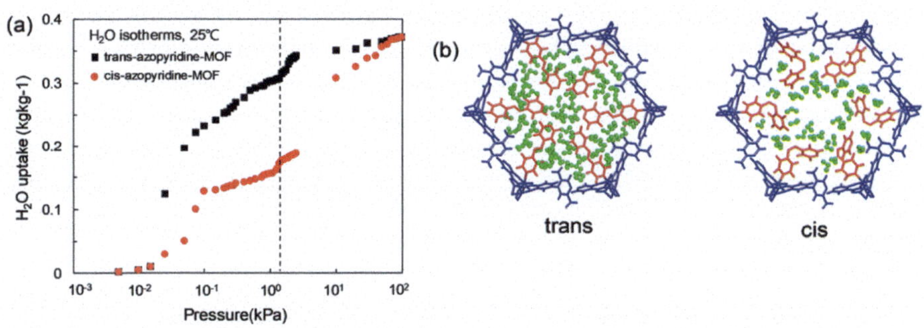

Fig. 4.4 **a** Azopyridine-water IRMOF74-III's absorption characteristics at 25 °C, and **b** Water adsorption structure at 2 kPa in both the trans and cis directions (Reprinted with permission from Suh et al. (2019) Copyright 2019 American Chemical Society)

per 1 g of the sample. Due to extremely hydrophilic groups on their pore surfaces, it was observed that the isotherm lines of MIL-101-NH_2 and MIL-101-SO_3H begin at lower pressure regions than those of pure MIL-101. However, due to the decreased hydrophilicity of the NO_2 group, identical water uptake isotherms were seen in MIL-101-NO_2 and pure MIL-101. It's interesting to note that all MIL-101 materials allow the release of the adsorbed water molecule at low temperatures (around 353 K). The hydrophilic environment inside the pores caused strong contact with water-molecules, and the presence of substituents played a significant role in modulating this interaction.

In the other studies by Ko and co-workers (2015) demonstrated the water vapour adsorption isotherms of zirconium based MOFs, UiO-67 and UiO-67-(NH_2). The UiO-67 material showed to adsorb little water at lower pressure up to 0.5 and a steep water uptake was observed at $P/P_0 = 0.6$ making a sigmoidal shape. The material was capable to adsorb 293 mg of water per 1 g. On the other hand, the water adsorption properties of UiO-67-(NH_2) showed an initial water uptake at lower pressure, $P/P_0 = 0$, started to increase from relative pressure of 0.2 (Ko et al. 2015). This was followed by water adsorption capacity of 173 mg g^{-1} at $P/P_0 = 0.3$. The high adsorption properties was due to the strong affinity of MOF toward water made by the presence of NH_2 group within the pores.

Jeremias et al. (2012) presented the use of MIL-100(Fe, Al) for water harvesting. Their observed water adsorption isotherm of MIL-100(Fe) was in a good agreement with the one reported by Kusgens et al. (2009). The water adsorption capacities at 25 °C which were obtained to be 0.75 and 0.50 g g^{-1} for Fe and Al based MIL-100, respectively. There was little bit drop of water adsorption at 40 °C for MIL-100(Fe), while MIL-100(Al) had the same water uptake at both temperatures. The difference between the two isotherms is mostly due to MIL-100(Al)'s lower water uptake than MIL-100 (Fe). Their findings were explained by the occurrence of adsorption and cluster formation at the hydrophilic metal sites at lower relative pressures ($P/P_0 < 0.25$). After that, the sequential filling of the 25 A and 29 A mesopores was what caused the sharp rise at $0.25 < P/P_0 < 0.45$. Likewise, the desorption properties of MIL-100(Fe) showed a hysteresis which was expected for mesoporous materials. On the other hand, MIL-100(Al) as an isostructure of MIL-100(Fe), it showed a similar shape of the isotherm with a hysteresis of about 30% smaller than that of MIL-100(Fe). When the relative pressure was < 0.2, the adsorption isotherms of the two materials were similar with the difference in the pore filling region. This was sounded by N_2 sorption for determination of total micropore volumes, which were observed to 0.87 and 0.65 cm^3 g^{-1} for Fe and Al based MIL-100, respectively. Hence there was a smaller water adsorption property of MIL-100(Al) as compared to MIL-100(Fe).

Rieth et al. (2019) prepared MOF from the ligand bis(1H-1,2,3-triazolo[4,5-b],[4′,5′-i])dibenzo[1,4]dioxin (H_2BTDD) and metal salts to form $Co_2(Cl)_2BTDD$ and $Ni_2(Cl)_2BTDD$. The Co2(Cl)$_2$BTDD material demonstrated a roughly 1 g of H_2O per 1 g of MOF capacity. The water isotherm for $Ni_2(OH)_2BTDD$ revealed a stronger initial hydrophilicity, with improved water absorption before 5% RH. The anion exchange was used on $Ni_2(Cl)_2BTDD$. However, the hydroxide material

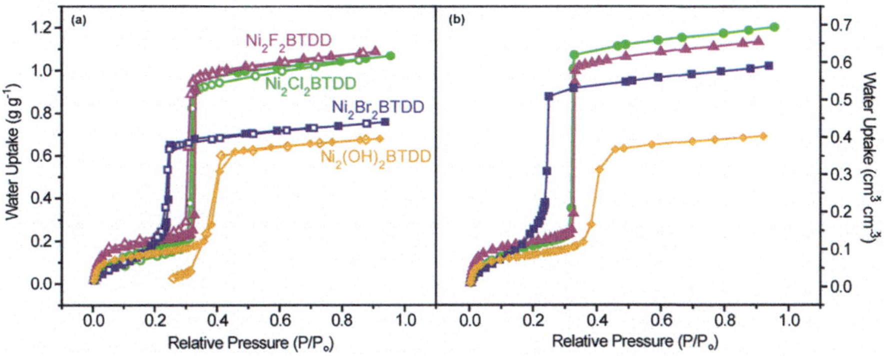

Fig. 4.5 **a** The water adsorption and desorption characteristics of Ni_2Br_2BTDD, Ni_2Cl_2BTDD, Ni_2F_2BTDD, and $Ni_2(OH)_2BTDD$ were measured at 25 °C, and **b** volumetric units for adsorption capabilities (Rieth et al. 2019)

showed a reduction in water capacity relative to the chloride equivalent at 5% RH. Additionally, as shown in Fig. 4.5, the OH group-based material shown a reduction in the water uptake step in the pore-filling area. Additionally, it was noted that this OH^- material had a rapid capacity decline with increasing the adsorption–desorption cycle as well as an irreversible desorption isotherm (Rieth et al. 2019). On the other hand, the anion exchange of Cl^- by F^- and Br^- was utilized to enhance $Ni_2(Cl)_2BTDD$'s water absorption. The Ni_2Br_2BTDD material's water isotherm revealed that the pore-filling step changed from 32% RH at 25 °C to 24% RH for the Br analogue (Fig. 4.5). As a result, Ni_2Br_2BTDD absorbed 0.64 g of water for every 1 g of MOF at RH levels below 25%. Due to a stronger hydrogen bonding contact being established between the water molecule and the pores of the Ni_2Br_2BTDD MOF than either the chloride or fluoride derivatives, the water absorption was less for this material.

The water absorption of the five distinct MOFs/$CaCl_2$ composites was reported by Shi et al. (). They measured how much water was adsorbing onto the bulk of the dry materials at any given time at 30 °C and 32% RH. Figure 4.6a shows the outcomes of these composites' water adsorption isotherms. It was demonstrated that very hydrophilic groups (–SO_3H and –$NH_2$2019) on the pore surfaces of hydrophilic MOF-based composites (MIL-101(Cr)-SO_3H/$CaCl_2$ and MIL-101(Cr)-NH_2/$CaCl_2$) increased water adsorption rates and volumes at low relative pressures. Additionally, the MIL-101(Cr)-SO_3H/$CaCl_2$ composite absorbs 0.6 g of H_2O for every 1 g of MOF, which is a notable improvement above the MIL-101(Cr)/CaCl2 composite without functionalization (0.47 g (H_2O)/g (MOF)). The water absorption of the MIL-101(Cr)-CH_3/$CaCl_2$ and MIL-101(Cr)-F/$CaCl_2$ materials, on the other hand, was reduced due to the presence of –CH_3 and –F hydrophobic groups (0.44 g (H_2O)/g (MOF), respectively. This discovery matched the XRD patterns seen in Fig. 4.6b. Table 4.1 lists the advantages and disadvantages of MOFs-based adsorbents with remarkable adsorption capabilities.

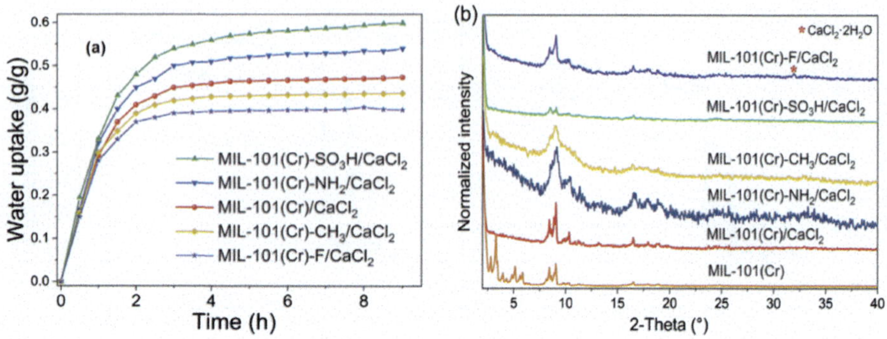

Fig. 4.6 **a** Water adsorption properties at 30 °C and RH = 32% and **b** XRD patterns of the five MOFs/$CaCl_2$ composites (Reprinted with permission from (Shi et al. 2019) Copyright 2019 Elsevier)

4.4 Conclusions and Future Trends in MOFs-Based Adsorbent

In this light, MOFs AWH technology is a novel method to supply clean water in every environment at any time of year while also disseminating and transportable water delivery. After overcoming the constraints of cost, scalability, and water collection for drinking, MOFs and MOF-based devices will progressively enter the market and considerably alleviate global water stress. To make AWH development a reality, we attempted to create MOFs with improved water sorption capabilities, perhaps satisfying the demands of a specific community (industrial scale). More study is required to compare the nanoporous AWH material to the micro-or traditional porous material. In addition to focusing on the presence of potent adsorbers in materials with pore sizes larger than the critical diameter of the desired adsorbate, this important study also pays attention to the cleanliness of the water and the amount of AWH.

Air pollutants, particles, and bacteria that adhere to the surface of the AWH material or that penetrate the interior are difficult to eliminate automatically due to their size or insolubility in water. We require a thorough assessment of the costs related to materials, catchment devices, and support/maintenance in order to bridge this research gap. A standard unit of measurement, such as the investigation-water production ratio per unit mass of material, payback period, or daily water production per amount spent, should serve as the foundation for the grading system, for instance.

Table 4.1 Summary of MOFs-based adsorbent with outstanding adsorption performance and challenges

Adsorbent ID	Climatic conditions	Adsorption–desorption rate	Challenges	Reference
MOF-841, $[Zr_6O_4(OH)_4(\text{methane-tribenzoate})_2(\text{formate})_4(H_2O)_2]$	Desert	Over 1 L/kg of MOF per day (about 600 cm^3/g)	The lower thermal conductivity of MOFs limit absorption and scalability	Xu and Yaghi (2020)
MOF-801, $[Zr_6O_4(OH)_4(\text{formate})_6]$	Desert	About 400 cm^3/g	Non-polar organic linkers and polar SBUs reduce the water-binding energy to the pores	Xu and Yaghi (2020)
MOF-801, $[Zr_6O_4(OH)_4(\text{formate})_6]$	Ambient	2.8 L/kg of MOF per day	Crystal diameter and porosity	Kim et al. (2017)
Mesoporous MOF	Simulated desert conditions	0.82 g of water per g of MOF	Need for scalability	Rieth et al. 2017)
MOF-801/Graphite	Desert	100 g of water per kg of MOF per day	Reduced direct heating utilizing solar thermal energy due to poor thermal conductivity, low infrared (IR) and near-IR absorption, and high heat capacity	Fathieh et al. (2018)

(continued)

Table 4.1 (continued)

Adsorbent ID	Climatic conditions	Adsorption–desorption rate	Challenges	Reference
Al-based MOF-303/Graphite	Desert	175 g of water per kg of MOF per day	Flexibility of design and industrial practical applications need to be addressed	Fathieh et al. (2018)
In-MIL-68	Room temperature	0.32 g of water per g of MOF	Critical challenges include low pore morphology and 1D channel structure. At low pressure, isotherms are convex, demonstrating the increasing influence of fluid–fluid interactions prior to an inflection point	Canivet et al. (2014b)
Ti_3C_2/UiO-66-NH_2	20% RH in a simulated indoor arid environment	57.8 mL of water per kg of MOF per hour	The adsorption and desorption kinetics are largely dependent on the light irradiation capacity and morphology ($\geq$ 103 mW/cm^2 and vertically aligned porous networks)	Wu et al. (2021)

Acknowledgements The authors gratefully acknowledge funding from South Africa's National Research Foundation (NRF) No. 116679 and all the other affiliated organizations.

Declaration of Competing Interest The authors disclose that they have no financially competing interests.

References

Akiyama G, Matsuda R, Sato H, Hori A, Takata M, Kitagawa S (2012) Effect of functional groups in MIL-101 on water sorption behavior. Micropor Mesopor Mater 157:89–93

Alnaser WE, Barakat A (2000) Use of condensed water vapour from the atmosphere for irrigation in Bahrain. Appl Energy 65(1–4):3–18

Baier W (1966) Studies on dew formation under semi-arid conditions. Agric Meteorol 3(1–2):103–112

Beysens D, Clus O, Mileta M, Milimouk I, Muselli M, Nikolayev VS (2007) Collecting dew as a water source on small islands: the dew equipment for water project in Bis˘ evo (Croatia). Energy 32(6):1032–1037

Bilal M, Sultan M, Morosuk T, Den W, Sajjad U, Aslam MM, Shahzad MW, Farooq M (2022) Adsorption-based atmospheric water harvesting: a review of adsorbents and systems. Int Commun Heat Mass Transfer 133:105961

Boretti A, Rosa L (2019) Reassessing the projections of the world water development report. NPJ Clean Water 2(1):1–6

Brown PS, Bhushan B (2016) Bioinspired materials for water supply and management: water collection, water purification and separation of water from oil. Philos Trans Royal Soc Math Phys Eng Sci 374(2073):20160135

Calderón M, Cereceda P, Larrain H, Osses P, Pérez L, Ibáñez M (2010) Alto Patache fog oasis in the Atacama Desert: geographical basis for a sustainable development program. In: Proceedings of the 5th international conference on fog, fog collection and dew, pp 25–30

Canivet J, Fateeva A, Guo Y, Coasne B, Farrusseng D (2014a) Water adsorption in MOFs: fundamentals and applications. Chem Soc Rev 43(16):5594–5617

Canivet J, Bonnefoy J, Daniel C, Legrand A, Coasne B, Farrusseng D (2014b) Structure–property relationships of water adsorption in metal–organic frameworks. New J Chem 38(7):3102–3111

Carvajal D, Minonzio JG, Casanga E, Muñoz J, Aracena A, Montecinos S, Beysens D (2018) Roof-integrated dew water harvesting in Combarbalá, Chile. J Water Supply Res Technol—AQUA 67(4):357–374

Domen JK, Stringfellow WT, Camarillo MK, Gulati S (2014) Fog water as an alternative and sustainable water resource. Clean Technol Environ Policy 16(2):235–249

Ejeian M, Wang RZ (2021) Adsorption-based atmospheric water harvesting. Joule 5(7):1678–1703

Fathieh F, Kalmutzki MJ, Kapustin EA, Waller PJ, Yang J, Yaghi OM (2018) Practical water production from desert air. Sci Adv 4(6):p.eaat3198

Furukawa H, Gandara F, Zhang YB, Jiang J, Queen WL, Hudson MR, Yaghi OM (2014) Water adsorption in porous metal–organic frameworks and related materials. J Am Chem Soc 136(11):4369–4381

Gad HE, Hamed AM, El-Sharkawy II (2001) Application of a solar desiccant/collector system for water recovery from atmospheric air. Renew Energy 22(4):541–556

Gado MG, Nasser M, Hassan AA, Hassan H (2022) Adsorption-based atmospheric water harvesting powered by solar energy: Comprehensive review on desiccant materials and systems. Process Saf Environ Prot 160:166–183

Garrod RP, Harris LG, Schofield WCE, McGettrick J, Ward LJ, Teare DOH, Badyal JPS (2007) Mimicking a Stenocara Beetle's back for microcondensation using plasmachemical patterned superhydrophobic—superhydrophilic surfaces. Langmuir 23(2):689–693

Gido B, Friedler E, Broday DM (2016) Assessment of atmospheric moisture harvesting by direct cooling. Atmos Res 182:156–162

Hanikel N, Pei X, Chheda S, Lyu H, Jeong W, Sauer J, Gagliardi L, Yaghi OM (2021) Evolution of water structures in metal-organic frameworks for improved atmospheric water harvesting. Science 374(6566):454–459

Hoskins BF, Robson R (1989) Infinite polymeric frameworks consisting of three dimensionally linked rod-like segments. J Am Chem Soc 111(15):5962–5964

Hu Y, Fang Z, Wan X, Ma X, Wang S, Fan S, Dong M, Ye Z, Peng X (2022) Carbon nanotubes decorated hollow metal–organic frameworks for efficient solar-driven atmospheric water harvesting. Chem Eng J 430:133086

Inbar O, Gozlan I, Ratner S, Aviv Y, Sirota R, Avisar D (2020) Producing safe drinking water using an atmospheric water generator (AWG) in an urban environment. Water 12(10):2940

Jabalia N, Kumar A, Kumar V, Rani R (2021) In Silico approach in drug design and drug discovery: an update. In: Innovations and implementations of computer aided drug discovery strategies in rational drug design. Springer, Singapore, pp 245–271

Jaen MVM (2002) Fog water collection in a rural park in the Canary Islands (Spain). Atmos Res 64(1–4):239–250

Jalali S, Aliabadi M, Mahdavinejad M (2021) Learning from plants: a new framework to approach water-harvesting design concepts. Int J Build Pathol Adapt

Jeremias F, Khutia A, Henninger SK, Janiak C (2012) MIL-100 (Al, Fe) as water adsorbents for heat transformation purposes—a promising application. J Mater Chem 22(20):10148–10151

Kabeel AE (2007) Water production from air using multi-shelves solar glass pyramid system. Renew Energy 32(1):157–172

Khalil B, Adamowski J, Shabbir A, Jang C, Rojas M, Reilly K, Ozga-Zielinski B (2016) A review: dew water collection from radiative passive collectors to recent developments of active collectors. Sustain Water Res Manag 2(1):71–86

Kim H, Yang S, Rao SR, Narayanan S, Kapustin EA, Furukawa H, Umans AS, Yaghi OM, Wang EN (2017) Water harvesting from air with metal-organic frameworks powered by natural sunlight. Science 356(6336):430–434

Kim H, Rao SR, Kapustin EA, Zhao L, Yang S, Yaghi OM, Wang EN (2018) Adsorption-based atmospheric water harvesting device for arid climates. Nat Commun 9(1):1–8

Kim H, Rao SR, LaPotin A, Lee S, Wang EN (2020) Thermodynamic analysis and optimization of adsorption-based atmospheric water harvesting. Int J Heat Mass Transf 161:120253

Ko N, Hong J, Sung S, Cordova KE, Park HJ, Yang JK, Kim J (2015) A significant enhancement of water vapour uptake at low pressure by amine-functionalization of UiO-67. Dalton Trans 44(5):2047–2051

Kondo M, Yoshitomi T, Matsuzaka H, Kitagawa S, Seki K (1997) Three-dimensional framework with channeling cavities for small molecules:{[M2 (4, 4′-bpy) 3 (NO3) 4]·xH2O} n (M=Co, Ni, Zn). Angew Chem Int Ed Engl 36(16):1725–1727

Kumar M, Yadav A (2015) Experimental investigation of design parameters of solar glass desiccant box type system for water production from atmospheric air. J Renew Sustain Energy 7(3):033122

Küsgens P, Rose M, Senkovska I, Fröde H, Henschel A, Siegle S, Kaskel S (2009) Characterization of metal-organic frameworks by water adsorption. Micropor Mesopor Mater 120(3):325–330

LaPotin A, Kim H, Rao SR, Wang EN (2019) Adsorption-based atmospheric water harvesting: impact of material and component properties on system-level performance. Acc Chem Res 52(6):1588–1597

Larrain H, Velásquez F, Pinto R, Lázaro P, Cereceda P, Osses P, Schemenauer RS (2001) Two years of fog measurements at the site "Falda Verde", north of Chañaral (Chile). In: Second international conference on fog and fog collection, pp 223–226

Larrain H, Velásquez F, Cereceda P, Espejo R, Pinto R, Osses P, Schemenauer RS (2002) Fog measurements at the site "Falda Verde" north of Chañaral compared with other fog stations of Chile. Atmos Res 64(1–4):273–284

Lawson HD, Walton SP, Chan C (2021) Metal–organic frameworks for drug delivery: a design perspective. ACS Appl Mater Interfaces 13(6):7004–7020

Li H, Eddaoudi M, Groy TL, Yaghi OM (1998) Establishing microporosity in open metal—organic frameworks: gas sorption isotherms for Zn (BDC)(BDC = 1, 4-benzenedicarboxylate). J Am Chem Soc 120(33):8571–8572

Li A, Bueno-Perez R, Wiggin S, Fairen-Jimenez D (2020) Enabling efficient exploration of metal–organic frameworks in the Cambridge structural database. CrystEngComm 22(43):7152–7161

Liu X, Beysens D, Bourouina T (2022) Water harvesting from air: current passive approaches and outlook. ACS Mater Lett 4(5):1003–1024

Llewellyn PL, Bourrelly S, Serre C, Vimont A, Daturi M, Hamon L, De Weireld G, Chang JS, Hong DY, Kyu Hwang Y, Hwa Jhung S (2008) High uptakes of CO2 and CH4 in mesoporous metal-organic frameworks mil-100 and mil-101. Langmuir 24(14):7245–7250

Mabokela TE, Somo TR, Maponya TC, Hato MJ, Makhado E, Makgopa K, Modibane KD (2022) Dynamic carbon dioxide uptake capacity of metal organic framework using thermogravimetrical evaluation at different CO2 pressure. Mater Lett 317:132086

Nioras D, Ellinas K, Constantoudis V, Gogolides E (2021) How different are fog collection and dew water harvesting on surfaces with different wetting behaviors? ACS Appl Mater Interfaces 13(40):48322–48332

Nørgaard T, Dacke M (2010) Fog-basking behaviour and water collection efficiency in Namib Desert Darkling beetles. Front Zool 7(1):1–8

Petersen JF, Sack D, Gabler RE (2016) Physical geography. Cengage Learn

Rieth AJ, Yang S, Wang EN, Dincă, M. (2017) Record atmospheric fresh water capture and heat transfer with a material operating at the water uptake reversibility limit. ACS Cent Sci 3(6):668–672

Rieth AJ, Wright AM, Skorupskii G, Mancuso JL, Hendon CH, Dincă, M. (2019) Record-setting sorbents for reversible water uptake by systematic anion exchanges in metal–organic frameworks. J Am Chem Soc 141(35):13858–13866

Rojas F, Carter V, Rosato M (2014) Fog collection technology transfer and co-creation projects in Falda Verde, Chile and Tojquia, Guatemala. In: Technologies for sustainable development. Springer, Cham, pp 275–286

Schemenauer RS, Zanetta N, Rosato M, Carter Gamberini MV (2016) The Tojquia, Guatemala fog collection project 2006 to 2016. In: 7th international conference on fog, fog collection and dew

Sharan G (2011) Harvesting dew with radiation cooled condensers to supplement drinking water supply in semi-arid. Int J Serv Learn Eng Humanitarian Eng Soc Entrepreneurship 6(1):130–150

Sharan G, Roy AK, Royon L, Mongruel A, Beysens D (2017) Dew plant for bottling water. J Clean Prod 155:83–92

Shi W, Zhu Y, Shen C, Shi J, Xu G, Xiao X, Cao R (2019) Water sorption properties of functionalized MIL-101 (Cr)-X (X=–NH2,–SO3H, H,–CH3,–F) based composites as thermochemical heat storage materials. Micropor Mesopor Mater 285:129–136

Suh BL, Chong S, Kim J (2019) Photochemically induced water harvesting in metal–organic framework. ACS Sustain Chem Eng 7(19):15854–15859

Suvindran N, Li F, Pan Y, Zhao X (2018) Characterization and bioreplication of tradescantia pallida inspired biomimetic superwettability for dual way patterned water harvesting. Adv Mater Interfaces 5(19):1800723

Teo HWB, Chakraborty A (2017) Water adsorption on various metal organic framework. In: IOP conference series: materials science and engineering, vol 272, No. 1. IOP Publishing, p 012019

Tu Y, Wang R, Zhang Y, Wang J (2018) Progress and expectation of atmospheric water harvesting. Joule 2(8):1452–1475

Unicef (2020) Water and the global climate crisis: 10 things you should know. https://www.unicef.org/stories/water-and-climate-change (2021). Accessed 29 April 2022

Wang JY, Wang RZ, Wang LW (2016) Water vapor sorption performance of ACF-CaCl2 and silica gel-CaCl2 composite adsorbents. Appl Therm Eng 100:893–901
Wang JY, Wang RZ, Tu YD, Wang LW (2018) Universal scalable sorption-based atmosphere water harvesting. Energy 165:387–395
Wu Q, Su W, Li Q, Tao Y, Li H (2021) Enabling continuous and improved solar-driven atmospheric water harvesting with Ti3C2-incorporated metal-organic framework monoliths. ACS Appl Mater Interfaces 13(32):38906–38915
Xing Q, Pan Y, Hu Y, Wang L (2020) Review of the biomolecular modification of the metal-organ-framework. Front Chem 8:642
Xu W, Yaghi OM (2020) Metal–organic frameworks for water harvesting from air, anywhere, anytime. ACS Cent Sci 6(8):1348–1354
Yaghi OM, O'Keeffe M, Ockwig NW, Chae HK, Eddaoudi M, Kim J (2003) Reticular synthesis and the design of new materials. Nature 423(6941):705–714
Zafar F, Sharmin E (eds) (2016) Metal-organic frameworks. BoD–Books on Demand
Zhang P (2016) Adsorption and desorption isotherms. KE Group
Zhang X, Chen Z, Liu X, Hanna SL, Wang X, Taheri-Ledari R, Maleki A, Li P, Farha OK (2020) A historical overview of the activation and porosity of metal–organic frameworks. Chem Soc Rev 49(20):7406–7427
Zheng Y, Bai H, Huang Z, Tian X, Nie FQ, Zhao Y, Zhai J, Jiang L (2010) Directional water collection on wetted spider silk. Nature 463(7281):640–643
Zhou X, Lu H, Zhao F, Yu G (2020) Atmospheric water harvesting: a review of material and structural designs. ACS Mater Lett 2(7):671–684

Chapter 5
Solar Adsorption-Based Atmospheric Water Harvesting Systems: Materials and Technologies

Mohamed G. Gado, Mohamed Nasser, and Hamdy Hassan

5.1 Introduction

Atmospheric water harvesting is a burgeoning technology for supplying water for decentralized production. It is confirmed to be a feasible solution, notably for areas that lack access to water and electricity (Gado et al. 2022b). It is worth mentioning that the atmospheric air encompasses 12,900 km^3 of water, which outperforms the water content in rivers by 6-times (Mulchandani et al. 2022). In that regard, adsorbent-based technologies are mainly based on water vapor saturation using adsorbent material; then, it is heated to release that stored vapor. Afterward, the released water vapor is condensed on a cold surface (Tu et al. 2018).

M. G. Gado
Department of Chemical Science and Engineering, Tokyo Institute of Technology, Tokyo 152-8552, Japan
e-mail: mohamed.gado@ejust.edu.eg

M. G. Gado · M. Nasser · H. Hassan (✉)
Energy Resources Engineering Department, Egypt-Japan University of Science and Technology (E-JUST), New Borg El-Arab City, Alexandria 21934, Egypt
e-mail: hamdy.aboali@ejust.edu.eg

M. Nasser
e-mail: mohamed.nafea@ejust.edu.eg

M. G. Gado
Mechanical Power Engineering Department, Faculty of Engineering at El-Mattaria, Helwan University, Cairo 11718, Egypt

M. Nasser
Mechanical Power Engineering Department, Faculty of Engineering, Zagazig University, Zagazig 44519, Egypt

H. Hassan
Mechanical Engineering Department, Faculty of Engineering, Assiut University, Assiut, Egypt

E. Fosso-Kankeu et al. (eds.), *Atmospheric Water Harvesting Development and Challenges*, Water Science and Technology Library 122,
https://doi.org/10.1007/978-3-031-21746-3_5

The workflow of atmospheric water harvesting includes catching water moisture from the atmospheric air; after that, it is condensed into a liquid. The water vapor is generally absorbed when its relative humidity is high, and its temperature is low. The momentum of the sorption is derived from the difference in vapor pressure between the surface of the desiccant material and the surrounding air. When the air is significantly dry, and the temperature is relatively high, additional energy is used to boost the vapor pressure of the desiccant surface to facilitate the desorption process. Finally, the desorbed vapor is condensed into a pure liquid water state. Throughout the desorption and condensing procedures, additional energy is required. The current chapter aims to illustrate the most recent techniques in materials and systems for air–water harvesting technologies.

5.1.1 AWH Working Concept

Figure 5.1 depicts a water harvesting concept comprised of two processes: sorption and release. The sorption process begins with adsorbing moisture from moist air at night. The collected water vapor is released from the saturated adsorbents during the daytime. The collection process, which consists of desorption and condensation, occurs during the day and lasts until the following cycle begins at the end of the day (Gado et al. 2022b).

Figure 5.2 indicates the potential application of adsorption-based AWH at lower RH. In that psychrometric chart, the green region reveals the potential application of direct cooling (condensation cycle) at higher RH (typically higher than 60%). Meanwhile, the pink region denotes the potential application of adsorption-based AWH, where there is limited RH. It can be highlighted that direct cooling at lower levels of moisture (Fig. 5.2a, b), the cooling down of thin air up to its dew point, requires large energy consumption. On the other hand, concentrating the water in the adsorbent material via exposing it to air directly (Fig. 5.2a, c) makes the cooling down process (Fig. 5.2c, d) less energy-intensive to attain its dew point. The potential for water harvesting is significant here, provided that the adsorbed water in the material's pores can be extracted without applying too much heat, making the operation efficient. However, finding a porous material capable of absorbing water at low RH

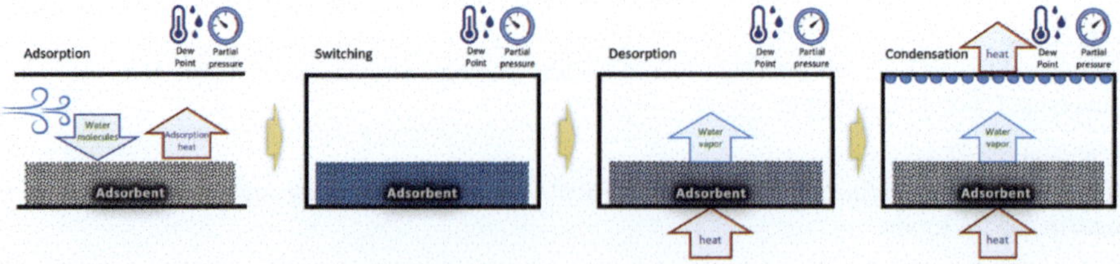

Fig. 5.1 Schematic of adsorption-based atmospheric water harvesting concept (Ejeian and Wang 2021)

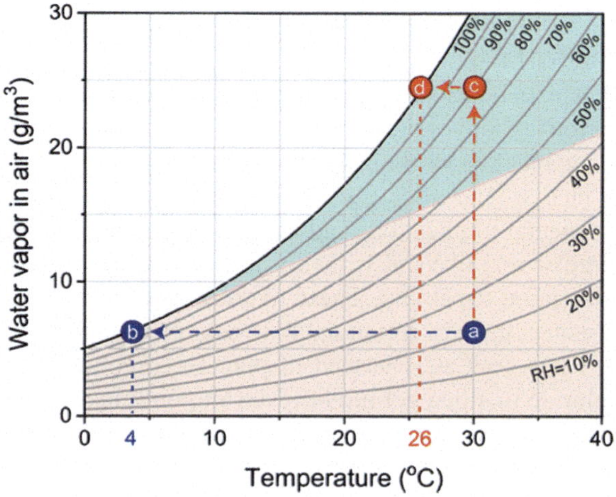

Fig. 5.2 Schematic diagram for the potential application of adsorption-based AWH and condensation-based AWH at lower RH and higher RH, respectively (Xu and Yaghi 2020)

and releasing it under practical conditions has been a major quest (Xu and Yaghi 2020).

5.2 Adsorption Materials for AWH

Several desiccant materials are utilized for atmospheric water harvesting. These desiccant materials are analyzed to forecast how much freshwater is yielded under different climatic conditions. This study covers the contemporary and next-generation adsorbents, like silica gel, expanded natural graphite, and activated carbon fibers, that are treated with other materials such as zeolite, sulfuric acid, and calcium chloride with various host materials, for instance, hydrogel and MOFs (Gado et al. 2022b; Hassan et al. 2022a).

Exceptional desiccant materials for AWH are described by Zhou et al. (2020):

- Greater adsorption capacity
- Smaller energy demand
- Fast kinetics
- Cycling stability.

5.2.1 Silica Gel

In water adsorption, silica gel proved its ability for large types of applications for water adsorption (Gado et al. 2022c, d, e). It shows a crucial part in surrounding air humidity adsorption when it possesses a low regenerative temperature and is affordable (Gado et al. 2022d); however, the low silica gel sorption capacity makes it has limited thermal stability and hardly desorbs water (Gado et al. 2021b; Hassan et al. 2022b). Moreover, the pure silica gel adsorption capacity cannot exceed 40 wt%. Hence, to boost adsorption capability, it is required to add additives. In this concern, various additives are mixed (e.g., hygroscopic salts and metal ions) with silica gel to augment its weaknesses. Different metal ions, including titanium, have been used to increase the silica gel sorption capacity; however, this method has negative environmental implications and is expensive (Okada et al. 2000; Hassan 2013). Adding hygroscopic like $CaCl_2$, LiBr, $MgSO_4$, and Ca $(NO_3)_2$ could increase the silica gel adsorption capability. These hygroscopic salts are non-toxic and inexpensive. For instance, the adsorption uptake could increase from 0.15 to 0.73 g/g due to the use of silica gel/$CaCl_2$ instead of pure silica gel at 30 °C. Moreover, adding calcium chloride to silica gel might result in pore blockage (Bu et al. 2013). At 30 °C, the water absorption of silica gel/LiBr might reach 0.6 g/g at a concentration of 53 wt% (Courbon et al. 2020).

5.2.2 *Zeolite*

Zeolite is alkaline earth metal-including crystal water or an aluminosilicate mineral of alkali (Yuan et al. 2016). Different zeolites, such as type Y, type X, and Type A, have been investigated. To promote the adsorption capacity of zeolites, hygroscopic salts like $CaCl_2$ have been added to decrease the required elevated regeneration temperatures (Chan et al. 2012). Several zeolites have been compared against metal–organic frameworks, revealing that the adsorption uptake of MIL-101(Cr) is superior to Zeolite 13X and AQSOA Z02 (Akiyama et al. 2012).

5.2.3 *Activated Carbon Fiber*

Activated carbon fiber (ACF) has superior adsorption kinetics and a Brunauer–Emmett–Teller (BET) surface area of 1380 m^2/g (Suzuki 1994). ACF is commonly utilized as a root adsorbent with host materials to increase its adsorption capacity. Different host materials are examined, such as calcium chloride ($CaCl_2$), lithium chloride (LiCl), and calcium chloride + magnesium sulfate (LiCl + $MgSO_4$). It was found ACF/$CaCl_2$ (1.7 kg/kg) has a superior uptake compared with ACF/LiCl (2.5 kg/kg). Also, it should be highlighted that LiCl is 10-folds more expensive than that $CaCl_2$.

5.2.4 Expanded Natural Graphite

Expanded natural graphite (ENG) is mixed with treated sulfuric acid (TSA) to increase the adsorption capability of the root adsorbent. For instance, the adsorption uptake of silica gel is augmented to 0.1 kg/kg in the case of using ENG-TSA as compared with 0.08 g/g for pure silica gel at ambient temperature and relative humidity of 30 °C and 70%, respectively (Zheng et al. 2014). Additionally, the ENG-TSA/silica gel is impregnated with LiCl aqueous solution, revealing an adsorption uptake of 0.65 kg/kg compared with 0.13 kg/kg for pure silica gel, ambient temperature, and relative humidity of 20 °C and 70%, respectively (Zheng and Wang 2019).

5.2.5 Metal–Organic Frameworks (MOFs)

MOFs have superior adsorption characteristics at different relative humidity. The most performant Metal–organic frameworks are described by fast response kinetics, high adsorption capacity at a lower temperature, cycling stability, non-toxic, and lower relative humidity (Mouchaham et al. 2020). Several MOFs have been examined, such as Zirconium-based as MOF-802 and MOF-801, MIL-series as MIL-125 and MIL-100, and extra classes (e.g., CAU-10-H and MOF-14). By screening twenty different types of MOFs, it was found that MOF-801 is remarkedly suitable for atmospheric water harvesting applications (Furukawa et al. 2014). MOF-801 could produce up to 0.25 L/kg in an arid climate (Kim et al. 2018). Under Arizona meteorological conditions, MOF-801 could generate 0.3–0.3 mL/kg/day at ambient temperature and relative humidity of 20 − 40 °C and 5 − 40%, respectively (Xu and Yaghi 2020). On the other hand, UiO-66 yielded about 0.04 kg/kg at ambient temperature and relative humidity of 25 °C and 40%, respectively (Trapani et al. 2016). Besides, it has been found that MIL-101(Cr) and MIL-100(Fe) have adsorption capacities of 1.5 kg/kg and 0.84 kg/kg at ambient temperature and relative humidity of 30 °C and 57%, respectively (Seo et al. 2012).

Moreover, MOFs have several drawbacks, such as high prices and low hydrothermal stability. Therefore, MOF-801 is introduced to curtail these drawbacks. Also, MOF-303 possesses favorable adsorption kinetics and lower cost than MOF-801, which attains a daily water production of 0.48 kg/kg during one complete cycle of 180 s (Fathieh et al. 2018). Using HKUST-1/LiCl composite can sustain 1.09 kg/kg at a relative humidity of 50% and ambient temperature of 25 °C (Zhao et al. 2020). Also, MIL-101(Cr))/LiCl could attain about 0.77 kg/kg at ambient temperature and relative humidity of 30 °C and 30%, respectively (Xu et al. 2020). It is found that when graphite oxide is used as a coating for MIL-101(Cr), the introduced material is suitable for low relative humidity conditions due to its excellent adsorption characteristics in dry conditions and its thermal stability.

5.2.6 *Hydrogel*

Employing a nanoporous super hygroscopic hydrogel has an absorption capacity of up to 420% of its nominal weight at lower driving temperatures up to 55 °C and is remarkably stable (i.e., 1000 absorption/desorption cycles) (Nandakumar et al. 2019). Likewise, adding hydrogel to Functionalized carbon nanotubes (FCNTs) is viable for increasing the absorption of the solar spectrum (Entezari et al. 2020). Besides, deliquescent salts, such as $CaCl_2$, are embedded with hydrogel to increase the adsorbed water quantities; however, dissolving in the adsorbed water. Consequently, it is proposed to use super porous hydrogel of acrylic acid and sodium acrylate (P(AA + SA)) along with its composite calcium chloride to ameliorate the adsorption uptake without dissolution (Mittal et al. 2021).

5.2.7 *Closure of the Adsorption Materials*

It is crucial to investigate the adsorption materials' swift response (kinetics) at different ambient and relative humidity conditions to select the most appropriate ones properly. Accordingly, the fast kinetics could share in increasing water productivity via using multi-cycling processes of adsorption and regeneration (Zhou et al. 2020). Low thermal conductivity is still a prominent drawback for the available adsorbent materials. Furthermore, using next-generation adsorbents has many barriers regarding the cost of synthesis. As demonstrated in Fig. 5.3, Feng et al. (2022) concluded the main barriers and prospects of various AWH materials. It is strongly anticipated that forthcoming research will focus on combining various AWH materials and using various diffusion mechanisms to attain higher water adsorption and desorption efficacies.

Table 5.1 highlights the key features of the adopted desiccant materials in the current chapter. It is noteworthy pointing out that zeolites are negatively operated using higher regeneration temperatures up to 300 °C. On the other hand, the regeneration process of MOFs is positively working at lower regeneration temperatures up to 50 °C.

5.3 Atmospheric Water Harvesting Technologies

5.3.1 *Technologies of AWH Devices Powered by Solar Energy*

The utilization of solar energy for different applications has been received considerable attention during the last decades, given its abundance specially at arid regions where is limited access to power (Gado et al. 2022a; Hassan et al. 2020). Various

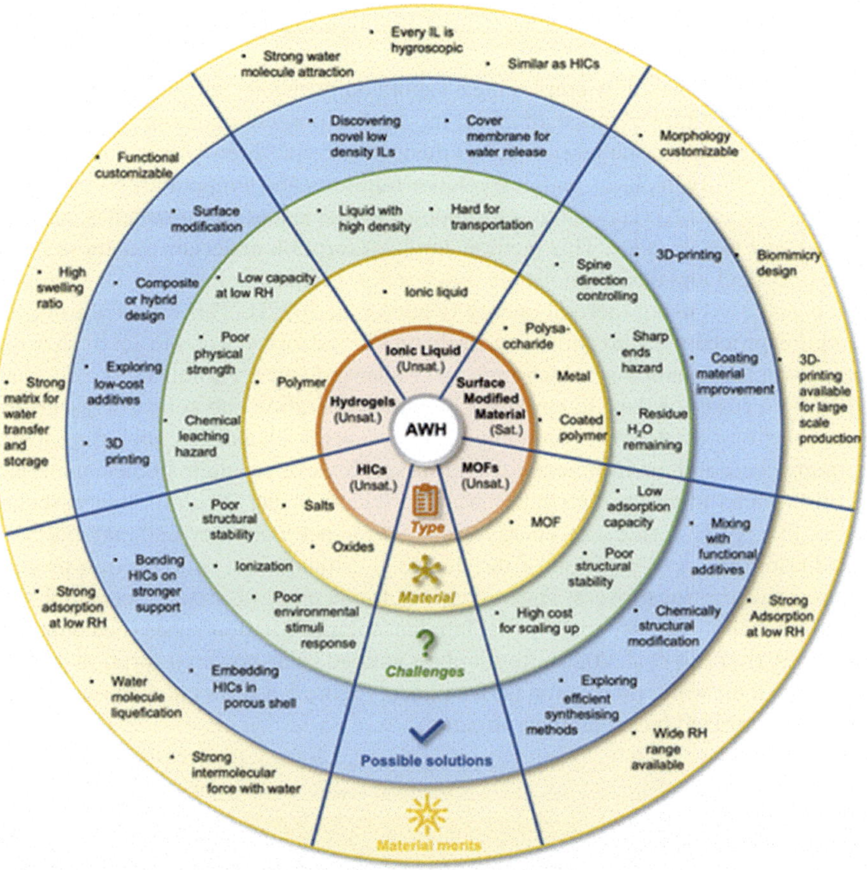

Fig. 5.3 Main barriers and prospects of AWH materials (Feng et al. 2022)

Table 5.1 Key specifications for different AWH materials

Material	Low regeneration temperatures	High adsorption uptake	Fast kinetics	Low RH operation
Zeolites	×	×	×	✓
Hydrogels and polymers	✓	✓	×	×
Salts	✓	×	×	✓
MOFs	✓	✓	✓	✓

AWH systems are used for water harvesting, highlighting their advantages and disadvantages. Active refrigeration AWH methods are the standard technique. They typically collect water vapor by employing a vapor compression cycle that uses power to lower the temperature of moist air under the air dew point temperature. The primary drawbacks of vapor compression AWH devices are their high-power consumption and inapplicability in places with low relative humidity and temperatures. Contrarily, the sorption-based AWH is another technology of AWH that has attracted academics over the past few decades. This method employs sorption materials like those listed above to collect and absorb moist air water vapor. Increasing the material's temperature is achieved using thermal energy (Gado et al. 2022f). This contributes to the desorption procedure's threshold and discharges the adsorbed water to get fresh water from adsorption materials, which may be condensed and accumulated (Gado et al. 2019). As a result of their low relative humidity operation and solar energy generation, sorption-based AWH technologies have several advantages, including lower operating costs and environmental friendliness. However, in their studies, materials scientists focus more on the cycle's water production rate than on other essential metrics like AWH device size, energy consumption per yield liter, and material cost. In addition, the cycle duration of the device is critical since a desorption process follows an adsorption process. The utilization of solar energy is extensively employed for different applications, notably for abundant solar irradiation zones (Nandakumar et al. 2019; Entezari et al. 2020). Researchers studied solar-powered sorption devices to improve their water production performance. Figure 5.4 shows how these systems may be categorized based on their structure.

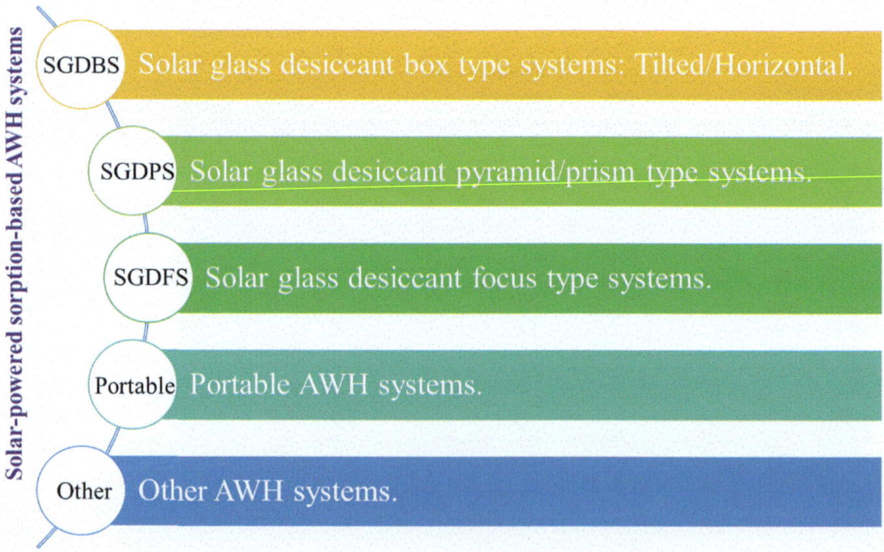

Fig. 5.4 Solar-powered AWH classification

5.3.2 *Solar Glass Desiccant Box-Type Systems (SGDBTS)*

5.3.2.1 Tilted SGDBTS

Desiccant material in the current configuration is placed at the base of the tilted box in wire mesh, as illustrated in Fig. 5.5. In this figure, the top of the box is composed of glass to allow solar radiation to flow through it during the daytime and in the box, functioning like a condenser to be used to condense the evaporated water. The box is tilted to collect condensed water in a bottle. For the desiccant material to absorb the air–water vapor, the humid air is permitted to enter the box overnight through the window. Water vapor condenses on the glass lid of the box and is collected in a bottle as a result of the thermal incident energy from the sun's rays enhancing the temperature. Saw wood and cotton cloth work as host material for the salt to create the desiccant pair (saw wood/$CaCl_2$) of the harvesting cycle.

Tilted SGDBTS can produce up to 200 mL of pure water when saw wood/$CaCl_2$ and silica-gel are utilized as a desiccant material (Kumar and Yadav 2015a, b). Another study is conducted in a dry climatic condition, Saudi Arabia, to show the device's performance (Abualhamayel and Gandhidasan 1997). The results revealed that it could produce about 1.92 kg/m^2 of pure water. Several modifications are conducted to the current configuration to analyze its performance under different conditions. The corrugated bed has a higher water production than the flat one, as mentioned in Gad et al. (2001). Additionally, the effect of adding a fan for air circulation (Gandhidasan and Abualhamayel 2010) and reflectors to the cover to maximize the gained solar radiation (Ji et al. 2007) are studied. The results showed that these modifications improved the device's performance. The impact of the tilt

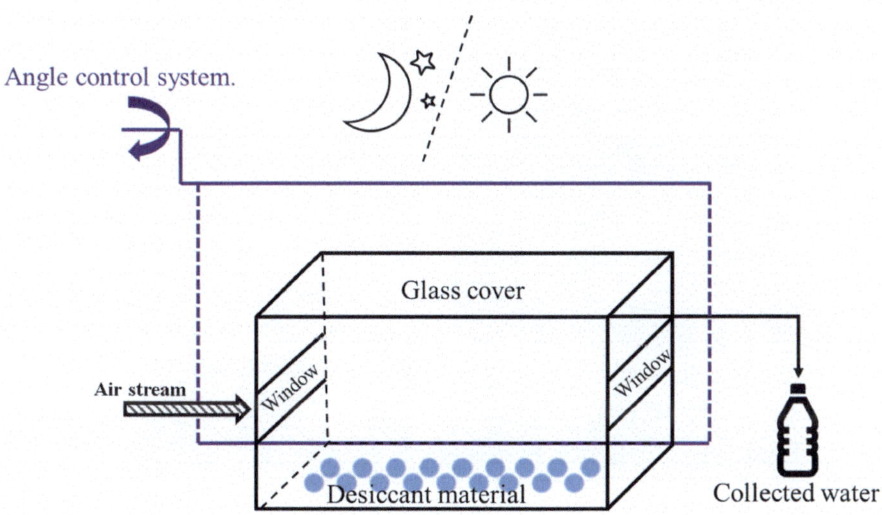

Fig. 5.5 Tilted SGDBS schematic diagram

angle of the box on water yield is investigated to estimate the optimum tilt angle (Kabeel 2006). The results of this study mentioned that the most significant amount of water produced is around 1.2 L/m^2 when the tilt angle equals 25°. The current configuration water yielded varies from 0.08 to 3.5 L/m^2 per kg of desiccant material.

5.3.2.2 Horizontal SGDBS

On the other hand, the current structure of SGDBS has a glass cover in a horizontal position facing the Earth's surface in which the tilt angle is equal to zero. Figure 5.6 shows the device components and work principle. The sorption unit includes desiccant material located inside the condenser unit. The desiccant material could be, for instance, silica gel (Sleiti et al. 2021) and MOFs (Suzuki 1994; Zheng et al. 2014). For humid nighttime air, either the upper cover may be removed, or a sidewall can be opened to allow adsorption of the water via the desiccant material. The glass cover is used throughout the day, such as a condenser. Occasionally, an aerogel coating is applied to the material throughout water yield to limit convective heat loss.

The current device can produce 0.25 L of pure water per kg of desiccant material, MOF-801 (Kim et al. 2018), and 0.16 L of purified water per kg of desiccant material, silica gel (Sleiti et al. 2021). The performance of different desiccant materials is tested at the same condition in the current configuration (Xu and Yaghi 2020). The conclusions of this test indicated that MOF-841 has the highest performance over MOF-801 and MOF-303. Yao et al. stated that this configuration could produce 25 L

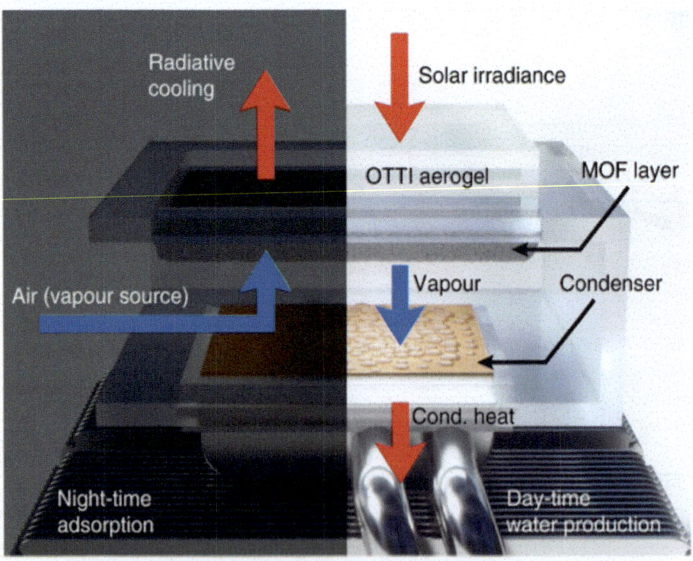

Fig. 5.6 Horizontal SGDBS working principle (Kim et al. 2018)

of pure water daily when the polyacrylate/graphene framework is used as a desiccant material (Yao et al. 2020).

Like Tilted SGDBS, many modifications are conducted to estimate the Horizontal SGDBS performance. The water production per day of the device is enhanced when the bed and the condenser are split, and water production reaches up to 2.5 L/m^2 per day (Qi et al. 2019). Furthermore, the water production increased up to 18% when multistage is applied instead of one stage of horizontal SGDBS (LaPotin et al. 2020). The current configuration yielded a freshwater production of 0.066–0.133 L/m^2 per kg desiccant material.

5.3.3 Solar Glass Desiccant Pyramid-Prism Type Systems (SGDPS)

The principal structure of the current configuration is multi-shelf beds in a trapezoidal or pyramid prism vessel, as demonstrated in Fig. 5.7. These shelves include desiccant material, and the principal function is to enhance bed surface area within the collector, hence increasing the amount of water collected. An inclining cover, which can be constructed of fiberglass or glass, is opened at night and closed during the day to allow humid air to pass through desiccant material and then condensate over an inclined cover during the daytime. Non-transparent material is used for the system's condenser head, which aids condensation more effectively.

A comparison is conducted with the SGDPS and SGDBS (horizontal) (Kabeel 2006) and SGDBS (tilted) (Gad et al. 2001). The results of this comparison stated that the water production of SGDPS is higher than these two systems by 90–95%.

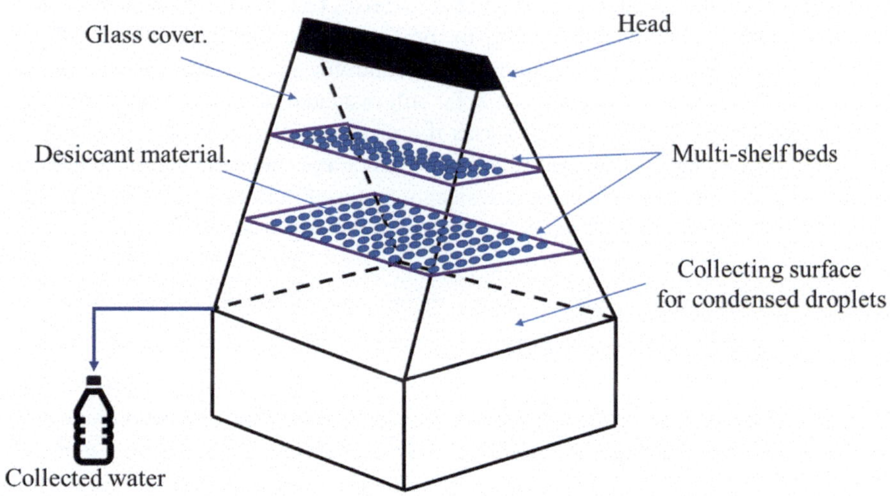

Fig. 5.7 SGDPS schematic diagram

Similarly, cloth/$CaCl_2$ is more suitable for this type because it has the most increased water production (Mohamed et al. 2017; Kabeel 2007). The system produces water of 1.06 L/m^2 per day using 22.96 MJ of solar energy per liter of water at the cost of \$0.055 (Elashmawy and Alatawi 2020). The AWH approach has shown excellent potential from low-humid locations up to 26.5% in the current investigation. Neither natural water supplies nor infrastructure is required to run this standalone device with promising performance. Remote and isolated tiny towns can benefit from the device's capabilities.

5.3.4 *Solar Glass Desiccant Focus Type System (SGDFTS)*

In the current AWH devices, the desiccant materials, like MOF or $CaCl_2$, are located inside a box representing a bed and a condenser. This box is placed in the parapolice trough or dish solar collector focus line/point. Concentrating the sun's reflected rays on the bed causes it to heat up and release its water vapor content. This concept is illustrated in Fig. 5.8. The performance is investigated under three desiccant materials. These desiccant materials are composite materials that contain LiCl, $CaCl_2$, and LiBr as adsorbents and sand as host materials (Srivastava and Yadav 2018). The results revealed that LiCl/sand could produce 90 ml of water per day, $CaCl_2$/sand produce115 ml daily, and LiBr/sand 73 ml of water per day. Furthermore, For $CaCl_2$/sand, LiCl/sand, and LiBr/sand, the system's average efficiency is 9.9%, 11.82%, and 11.10%, respectively. Regarding economic analysis, $CaCl_2$/sand, LiCl/sand, and LiBr/sand have yearly costs of \$0.71, \$0.53, and \$0.86 per liter of water generated from the atmosphere utilizing SGDFS.

Replacing the box with a double-slope half cylindrical basin solar distiller with four longitudinal fins increases the surface area (Essa et al. 2020). The results demonstrate that water production increased up to 166% due to this modification. In addition, a combination of SGDFS with a parabolic concentrator concentrates the sunlight on the desiccant material bed and tubular solar still instead of the box (Elashmawy and Alshammari 2020). This approach raises the production efficiency of water from 0.13 to 0.51 L per kg of desiccant material, nearly three times the increase in water productivity. As mentioned before, a desiccant material in a corrugated rolled shape improves water production compared to a flat one (Wang et al. 2017a, b).

5.3.5 *Portable AWH Systems*

Portable AWH systems, on the other hand, may be taken and set up anywhere, unlike the other systems described. A commercial device weighing 100 kg may generate 1450 g of water daily. As a result, these systems are recommended for catastrophe areas with intermittent electricity and no access to potable water.

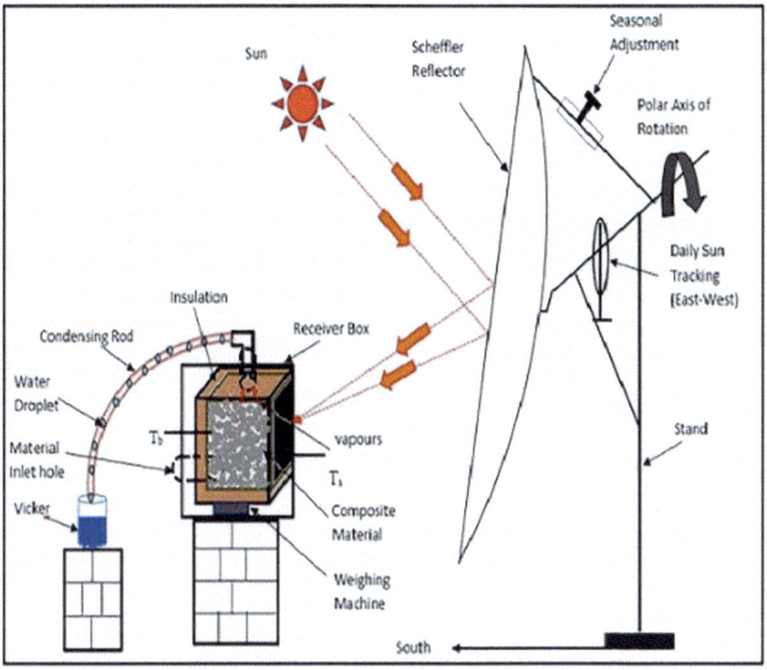

Fig. 5.8 SGDFS schematic diagram (Srivastava and Yadav 2018)

Figure 5.9a demonstrated an innovative AWH device created by Talaat et al. (2018). This device is constructed based on four portions. The absorber portion is a layer thin of cotton cloth/$CaCl_2$ in the shape of a right-angle equilateral triangle. A transparent cover represents the second part and consists of two conical united. The third part carries the absorber and cover, called the telescopic stick. The last part is a hose connected to a transparent cover base and a flask to collect the condensed water.

Furthermore, during the night, the desiccant solution-impregnated fabric layer of the conical absorber is exposed to ambient air, allowing the Calcium Chloride solution to soak up excess moisture in the humid air. In contrast, throughout the sunlight, the cover is closed. The temperature of the absorber increases due to solar energy, and the moisture in the absorber is evaporated and then condensed on the cover. The findings indicated that the maximum system efficiency was 22.56% when the water yield ranged from 0.3295–0.63 kg/m^2 per day.

Another innovation of AWH is presented by Fathy et al. (2020), as demonstrated in Fig. 5.9b. This device is foldable to easily carry and install anywhere that suffers from pure water shortage. The results mentioned that, during all tests, the range of production is 272–750 g/day.

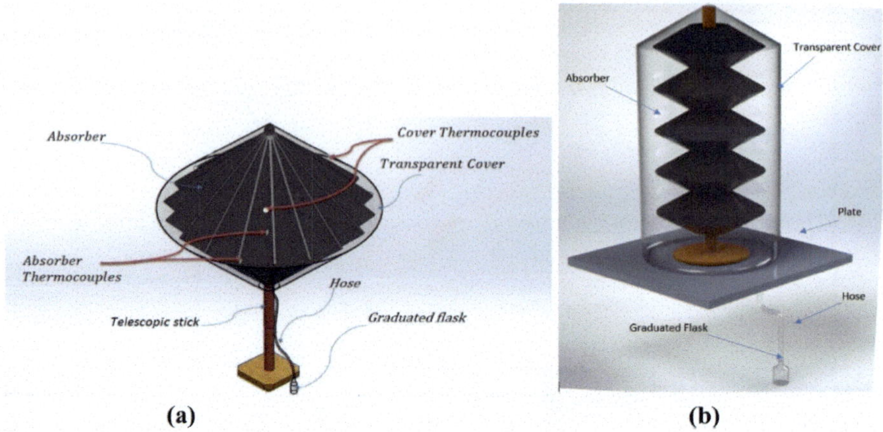

Fig. 5.9 Portable AWH systems (Talaat et al. 2018; Fathy et al. 2020)

5.3.6 Supplementary AWH Systems

Several operation principles illustrate different AWH devices in the current part. Water vapor adsorption and desorption may occur concurrently using an innovative cylindrical AWH unit (Li et al. 2020). Cylinder's bottom is exposed to the surroundings even though the sunlight strikes the top of the cylinder. Using a nanocarbon desiccant with LiCl in the cylinder base, the desiccant absorbs 100% of its weight in water from the surrounding air in under three hours. Still, the water vapor is swiftly out in the upper portion in 1/2 h in 1 kW/m^2 solar irradiation. This device can yield water of 1.6 kg for each 1 kg desiccant material in 10 h. These ten hours contained three harvesting/releasing cycles. Moreover, the HCS-LiCl nano adsorbent weighs 12.6 g in this study.

A nylon mesh bag including super moisture absorbent gel can work as an AWH with an efficiency of 81% and produces water at a rate of 3.9 kg per one kg of desiccant material (Zhao et al. 2019). It is necessary to take care and caution when dealing with these materials because some of them may be harmful to the produced water and make them dangerous and unsuitable for drinking (Yang et al. 2020). A dual adsorption/desorption device performs better than one cycle (Wang et al. 2021). The results revealed that the double cycle produces 0.42 kg of water per kilogram of desiccant material, while the monocycle produces 0.39 kg per kilogram of desiccant material.

To enhance water yield, an AWH prototype with adsorption/desorption cycles (see Fig. 5.10) was studied by Xu et al. (2021). The suggested system was shown to have a substantial water production of 2.12 L per kilogram of adsorbent.

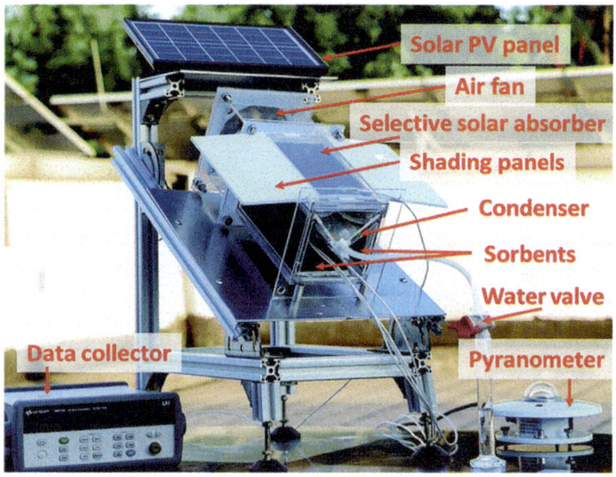

Fig. 5.10 Schematic diagram of the adopted prototype by Xu et al. (2021)

5.3.7 Closure of AWH Systems

Table 5.2 presents a detailed summary of the various technologies and host/desiccant materials for atmospheric water collecting. Numerous examination aspects, including the design of the proposed technology and the chosen host/desiccant material, have been examined to compare the thermal efficiency and financial viability of atmospheric water harvesting systems. As a result, various metrics have been evaluated to trade off the indicators of system viability. It should be noted that the current comparison examines several designs and materials under various climatic circumstances. In terms of water productivity, a very promising water production of 20 L/kg was found when metal–organic frameworks like MOF-801 and MOF-303 were used, demonstrating the potential use of these materials. The porous sodium polyacrylate/graphene framework (PGF) also beats the most effective desiccant materials, with PFG able to achieve water productivity of 25 L/kg.

Limitations and Prospects

Various characteristics and adsorption processes influence the development and promotion of desiccant compounds. Because of this, several obstacles prevent these materials from widespread use.

- **Uptakes of adsorption**: It is essential to use high water uptake desiccant materials to supply greater quantities of fresh water.
- **Reusability and cyclic stability**: Hydrothermal constancy of desiccant materials is insufficient for moisture transfer, and the multicyclic potential does not significantly reduce adsorption.
- **Scalability**: In terms of water productivity, scalability is the most crucial issue because there is a significant variation in the single-gram yield of desiccant materials and vast volumes.

Table 5.2 Solar-based AWH technologies

System	Adsorbent material	Water production	Reference
Horizontal-SGDBS (dual stage)	AQSOA Z01	1.48 L/m^2/kg	LaPotin et al. (2020)
Horizontal-SGDBS	MOF-801	2.80 L/kg	Kim et al. (2017)
Horizontal-SGDBS	MOF-801	0.25 L/kg	Kim et al. (2018)
Cylindrical solar still	Super moisture-absorbent gel (SMAG)	3.9 L/kg	Zhao et al. (2019)
Horizontal-SGDBS	MOF-801 MOF-303	20 L/kg	Xu and Yaghi (2020)
Horizontal-SGDBS	PGF	25 L/kg	Yao et al. (2020)
Cylindrical AWH	HCS-LiCl nano adsorbent	1.6 L/kg	Li et al. (2020)
Portable	Cotton cloth/$CaCl_2$	0.3295–0.631 L/m^2	Talaat et al. (2018)
Tilted-SGDBS	$CaCl_2$ solution	1.15–1.92 L/m^2	Gandhidasan and Abualhamayel (2010)
Tilted-SGDBS	Sand/$CaCl_2$	1.0 L/m^2	Hamed et al. (2011)
Horizontal-SGDBS	Hygroscopic ionic liquid	2.8 L/m^2	Qi et al. (2019)
Horizontal-SGDBS	MOF-303	0.7 L/m^2 (1.3 L/kg)	Hanikel et al. (2019)
Portable	Cotton cloth/$CaCl_2$	0.00064–0.0016 L/m^2	Fathy et al. (2020)
SGDFS	Silica gel	0.4 L/m^2	Essa et al. (2020)
Horizontal-SGDBS	Silica gel	0.8 L	Sleiti et al. (2021)

- **Safety**: Utilizing a filter of water to cleanse the gathered water avoids the harmful effects of desiccant compounds.

However, the utilization of AWH technologies is restricted by the following factors:

- There is a lack of large-scale practical deployment and long-term stability of AWH systems. Therefore, daily water yield is still not sufficient to meet the demand. So, there are still considerable challenges and problems to be solved before bringing scaled-up AWH systems to real-life personal, residential, commercial, or industrial applications.
- Improved condensation procedures might help improve the AWH system's overall thermal efficiency.
- Due to the direct interaction between desiccant materials and vapor and blown wind, collected water from AWH systems should be constantly monitored for dirt, metallic radicals, and dust.

The following factors are suggested based on the most recent research discussed earlier:

1. Developing innovative desiccant materials with high adsorption capabilities, rapid reaction times, and long-term stability to meet the everyday water needs of people throughout the world.
2. Creating and building large-scale AWH systems to improve water productivity so water harvesting can be used in real life.
3. Optimizing the temperature distribution during condensation improves the system's thermal efficiency, creating considerable freshwater.
4. Further water quality and filtration tests are necessary to determine whether the produced freshwater meets the World Health Organization (WHO) drinking water requirements.

5.4 Conclusions

Adsorption-based atmospheric water harvesting, as a promising technology for freshwater production, offers a wide range of intriguing uses for desalination, particularly in dry regions where water and power are unsatisfactory. The present chapter has thoroughly proven AWH technologies based on various desiccant materials. In light of the preceding discussion, the subsequent concluding points might be drawn as follows:

- Various AWH systems use composite desiccant materials that combine a host substance like ACF with hygroscopic salt or silica gel. This chapter demonstrates that certain materials are superior to others in collecting atmospheric moisture.
- Calcium chloride, as a composite material, performs admirably due to its low cost and lack of toxicity. Binary salts are also more water-absorbing than single salts in the AWH systems. However, LiCl is ten times more expensive than $CaCl_2$ in these materials.
- Hydrogel has long been regarded as a potential material for collecting water from the atmosphere because of its threefold weight capacity for water adsorption.
- MOFs' tailorable structures and chemical tunability have revolutionized water collecting applications, especially in dry environments with slight humidity. For instance, when MOF-303 and MOF-801 were used, a promising water output of 0.02 L/g was found to exist.
- Desiccant materials have several limitations restricting their water absorption, and new materials are needed to improve these shortcomings. Limited heat conductivity, lower water absorption as silica gel, expensive cost as MOFs, and lower sorption ability at lower relative humidity and high regeneration temperature are some of the downsides of this technology (e.g., polymers).
- In terms of freshwater production, the adsorption kinetics and capability and the architecture of the atmospheric water are vital parameters that control the production process.

- The water production from SGDBS can reach up to 3.5 L/m^2 per kg of desiccant material.
- The water production for SGDFTS can reach up to 0.51 L/h.
- Portable water harvesting devices are one of the most promising options for capturing water from the atmosphere. Solar-powered drinking water may be provided in dry areas because of its mobility and adaptability.

References

Abualhamayel HI, Gandhidasan P (1997) A method of obtaining fresh water from the humid atmosphere. Desalination 113:51–63. https://doi.org/10.1016/S0011-9164(97)00114-8

Akiyama G, Matsuda R, Sato H et al (2012) Effect of functional groups in MIL-101 on water sorption behavior. Micropor Mesopor Mater 157:89–93. https://doi.org/10.1016/j.micromeso.2012.01.015

Bu X, Wang L, Huang Y (2013) Effect of pore size on the performance of composite adsorbent. Adsorption 19:929–935. https://doi.org/10.1007/s10450-013-9513-8

Chan KC, Chao CYH, Bahrami M (2012) Heat and mass transfer characteristics of a zeolite 13X/$CaCl_2$ composite adsorbent in adsorption cooling systems. ASME 2012 6th international conference on energy sustainable ES 2012, collocated with ASME 2012 10th international conference on fuel cell science, engineering technology, pp 49–58. https://doi.org/10.1115/ES2012-91246

Courbon E, D'Ans P, Skrylnyk O, Frère M (2020) New prominent lithium bromide-based composites for thermal energy storage. J Energy Storage 32:101699. https://doi.org/10.1016/j.est.2020.101699

Ejeian M, Wang RZ (2021) Adsorption-based atmospheric water harvesting. Joule 5:1678–1703. https://doi.org/10.1016/j.joule.2021.04.005

Elashmawy M, Alatawi I (2020) Atmospheric water harvesting from low-humid regions of Hail City in Saudi Arabia. Nat Resour Res 29:3689–3700. https://doi.org/10.1007/S11053-020-09662-Y/TABLES/3

Elashmawy M, Alshammari F (2020) Atmospheric water harvesting from low humid regions using tubular solar still powered by a parabolic concentrator system. J Clean Prod 256:120329. https://doi.org/10.1016/j.jclepro.2020.120329

Entezari A, Ejeian M, Wang R (2020) Super atmospheric water harvesting hydrogel with alginate chains modified with binary salts. ACS Mater Lett 2:471–477. https://doi.org/10.1021/acsmaterialslett.9b00315

Essa FA, Elsheikh AH, Sathyamurthy R et al (2020) Extracting water content from the ambient air in a double-slope half-cylindrical basin solar still using silica gel under Egyptian conditions. Sustain Energy Technol Assess 39:100712. https://doi.org/10.1016/j.seta.2020.100712

Fathieh F, Kalmutzki MJ, Kapustin EA et al (2018) Practical water production from desert air. Sci Adv 4:1–10. https://doi.org/10.1126/sciadv.aat3198

Fathy MH, Awad MM, Zeidan ESB, Hamed AM (2020) Solar powered foldable apparatus for extracting water from atmospheric air. Renew Energy 162:1462–1489. https://doi.org/10.1016/j.renene.2020.07.020

Feng A, Akther N, Duan X et al (2022). Recent development of atmospheric water harvesting materials: a review. https://doi.org/10.1021/acsmaterialsau.2c00027

Furukawa H, Gándara F, Zhang YB et al (2014) Water adsorption in porous metal-organic frameworks and related materials. J Am Chem Soc 136:4369–4381. https://doi.org/10.1021/ja500330a

Gad HE, Hamed AM, El-Sharkawy II (2001) Application of a solar desiccant/collector system for water recovery from atmospheric air. Renew Energy 22:541–556. https://doi.org/10.1016/S0960-1481(00)00112-9

Gado M, Elgendy E, Elsayed K, Fatouh M (2019) Performance enhancement of an adsorption chiller by optimum cycle time allocation at different operating conditions. Adv Mech Eng 11:1–12. https://doi.org/10.1177/1687814019884780

Gado MG, Ookawara S, Nada S, El-Sharkawy II (2021a) Hybrid sorption-vapor compression cooling systems: a comprehensive overview. Renew Sustain Energy Rev 143:110912. https://doi.org/10.1016/j.rser.2021.110912

Gado MG, Megahed TF, Ookawara S et al (2021b) Performance and economic analysis of solar-powered adsorption-based hybrid cooling systems. Energy Convers Manag 238:114134. https://doi.org/10.1016/j.enconman.2021.114134

Gado MG, Ookawara S, Nada S et al (2022a) Performance assessment of photovoltaic/thermal (PVT) hybrid adsorption-vapor compression refrigeration system. J Energy Syst 6:209–220. https://doi.org/10.30521/jes.1002871

Gado MG, Nasser M, Hassan AA, Hassan H (2022b) Adsorption-based atmospheric water harvesting powered by solar energy: Comprehensive review on desiccant materials and systems. Process Saf Environ Prot. https://doi.org/10.1016/j.psep.2022.01.061

Gado MG, Nada S, Ookawara S, Hassan H (2022c) Energy management of standalone cascaded adsorption-compression refrigeration system using hybrid biomass-solar-wind energies. Energy Convers Manag 258:115387. https://doi.org/10.1016/j.enconman.2022.115387

Gado MG, Megahed TF, Ookawara S et al (2022d) Potential application of cascade adsorption-vapor compression refrigeration system powered by photovoltaic/thermal collectors. Appl Therm Eng 118075. https://doi.org/10.1016/j.applthermaleng.2022.118075

Gado MG, Ookawara S, Nada S, Hassan H (2022e) Renewable energy-based cascade adsorption-compression refrigeration system: Energy, exergy, exergoeconomic and enviroeconomic perspectives. Energy 124127. https://doi.org/10.1016/j.energy.2022.124127

Gado MG, Ookawara S, Nada S, Hassan H (2022f) Performance investigation of hybrid adsorption-compression refrigeration system accompanied with phase change materials—intermittent characteristics. Int J Refrig 142:66–81. https://doi.org/10.1016/j.ijrefrig.2022.06.007

Gandhidasan P, Abualhamayel HI (2010) Investigation of humidity harvest as an alternative water source in the Kingdom of Saudi Arabia. Water Environ J 24:282–292. https://doi.org/10.1111/j.1747-6593.2009.00189.x

Hamed AM, Aly AA, Zeidan E-SB (2011) Application of solar energy for recovery of water from atmospheric air in climatic zones of Saudi Arabia. Nat Resour 02:8–17. https://doi.org/10.4236/nr.2011.21002

Hanikel N, Prévot MS, Fathieh F et al (2019) Rapid cycling and exceptional yield in a metal-organic framework water harvester. ACS Cent Sci 5:1699–1706. https://doi.org/10.1021/acscentsci.9b00745

Hassan H (2013) An experimental work on the effect of injection molding parameters on the cavity pressure and product weight. Int J Adv Manuf Technol 67:675–686. https://doi.org/10.1007/s00170-012-4514-4

Hassan H, Yousef MS, Ahmed MS, Fathy M (2020) Energy, exergy, environmental, and economic analysis of natural and forced cooling of solar still with porous media. Environ Sci Pollut Res 27:38221–38240. https://doi.org/10.1007/s11356-020-09995-4

Hassan AA, Elwardany AE, Ookawara S, Hassan H (2022a) Performance and economic analysis of hybrid solar collectors-powered integrated adsorption/reverse osmosis multigeneration system. Int J Energy Res 1–24.https://doi.org/10.1002/er.8512

Hassan AA, Elwardany AE, Ookawara S, Hassan H (2022b) Energy, exergy, economic and environmental (4E) assessment of hybrid solar system powering adsorption-parallel/series ORC multigeneration system. Process Saf Environ Prot. https://doi.org/10.1016/J.PSEP.2022.06.024

Ji JG, Wang RZ, Li LX (2007) New composite adsorbent for solar-driven fresh water production from the atmosphere. Desalination 212:176–182. https://doi.org/10.1016/j.desal.2006.10.008

Kabeel AE (2006) Application of sandy bed solar collector system for water extraction from air. Int J Energy Res 30:381–394. https://doi.org/10.1002/er.1155

Kabeel AE (2007) Water production from air using multi-shelves solar glass pyramid system. Renew Energy 32:157–172. https://doi.org/10.1016/j.renene.2006.01.015

Kim H, Yang S, Rao SR, Narayanan S et al (2017) Water harvesting from air with metal-organic frameworks powered by natural sunlight. Science 434(80):430–434. https://doi.org/10.1126/science.aam8743

Kim H, Rao SR, Kapustin EA et al (2018) Adsorption-based atmospheric water harvesting device for arid climates. Nat Commun 9:1–8. https://doi.org/10.1038/s41467-018-03162-7

Kumar M, Yadav A (2015a) Experimental investigation of design parameters of solar glass desiccant box type system for water production from atmospheric air. J Renew Sustain Energy 7.https://doi.org/10.1063/1.4922142

Kumar M, Yadav A (2015b) Experimental investigation of solar powered water production from atmospheric air by using composite desiccant material "CaCl2/saw wood." Desalination 367:216–222. https://doi.org/10.1016/j.desal.2015.04.009

LaPotin A, Zhong Y, Zhang L et al (2020) Dual-stage atmospheric water harvesting device for scalable solar-driven water production. Joule 1–17.https://doi.org/10.1016/j.joule.2020.09.008

Li R, Shi Y, Wu M et al (2020) Improving atmospheric water production yield: enabling multiple water harvesting cycles with nano sorbent. Nano Energy 67:104255. https://doi.org/10.1016/j.nanoen.2019.104255

Mittal H, Al AA, Alhassan SM (2021) Hybrid super-porous hydrogel composites with high water vapor adsorption capacity—adsorption isotherm and kinetics studies. J Environ Chem Eng 9:106611. https://doi.org/10.1016/j.jece.2021.106611

Mohamed MH, William GE, Fatouh M (2017) Solar energy utilization in water production from humid air. Sol Energy 148:98–109. https://doi.org/10.1016/j.solener.2017.03.066

Mouchaham G, Cui FS, Nouar F et al (2020) Metal-organic frameworks and water: 'from old enemies to friends'? Trends Chem 2:990–1003. https://doi.org/10.1016/j.trechm.2020.09.004

Mulchandani A, Edberg J, Herckes P, Westerhoff P (2022) Seasonal atmospheric water harvesting yield and water quality using electric-powered desiccant and compressor dehumidifiers. Sci Total Environ 825:153966. https://doi.org/10.1016/j.scitotenv.2022.153966

Nandakumar DK, Zhang Y, Ravi SK et al (2019) Solar energy triggered clean water harvesting from humid air existing above sea surface enabled by a hydrogel with ultrahigh hygroscopicity. Adv Mater 31:1–7. https://doi.org/10.1002/adma.201806730

Okada K, Tomita T, Kameshima Y et al (2000) Effect of preparation conditions on the porous properties of coprecipitated Al2O3-SiO2 xerogels synthesized from aluminium nitrate nonahydrate and tetraethylorthosilicate. Micropor Mesopor Mater 37:355–364. https://doi.org/10.1016/S1387-1811(99)00280-2

Qi H, Wei T, Zhao W et al (2019) An interfacial solar-driven atmospheric water generator based on a liquid sorbent with simultaneous adsorption-desorption. Adv Mater 31:1–9. https://doi.org/10.1002/adma.201903378

Seo YK, Yoon JW, Lee JS et al (2012) Energy-efficient dehumidification over hierarchically porous metal-organic frameworks as advanced water adsorbents. Adv Mater 24:806–810. https://doi.org/10.1002/adma.201104084

Sleiti AK, Al-khawaja H, Al-khawaja H, Al-ali M (2021) Harvesting water from air using adsorption material—prototype and experimental results material from Institut Lavoisier. Sep Purif Technol 257:117921. https://doi.org/10.1016/j.seppur.2020.117921

Srivastava S, Yadav A (2018) Water generation from atmospheric air by using composite desiccant material through fixed focus concentrating solar thermal power. Sol Energy 169:302–315. https://doi.org/10.1016/j.solener.2018.03.089

Suzuki M (1994) Activated carbon fiber: fundamentals and applications. Carbon N Y 32:577–586. https://doi.org/10.1016/0008-6223(94)90075-2

Talaat MA, Awad MM, Zeidan EB, Hamed AM (2018) Solar-powered portable apparatus for extracting water from air using desiccant solution. Renew Energy 119:662–674. https://doi.org/10.1016/j.renene.2017.12.050

Trapani F, Polyzoidis A, Loebbecke S, Piscopo CG (2016) On the general water harvesting capability of metal-organic frameworks under well-defined climatic conditions. Micropor Mesopor Mater 230:20–24. https://doi.org/10.1016/j.micromeso.2016.04.040

Tu Y, Wang R, Zhang Y, Wang J (2018) Progress and expectation of atmospheric water harvesting. Joule 2:1452–1475. https://doi.org/10.1016/j.joule.2018.07.015

Wang JY, Liu JY, Wang RZ, Wang LW (2017a) Experimental investigation on two solar-driven sorption based devices to extract fresh water from atmosphere. Appl Therm Eng 127:1608–1616. https://doi.org/10.1016/j.applthermaleng.2017.09.063

Wang JY, Wang RZ, Wang LW, Liu JY (2017b) A high efficient semi-open system for fresh water production from atmosphere. Energy 138:542–551. https://doi.org/10.1016/j.energy.2017.07.106

Wang W, Pan Q, Xing Z et al (2021) Viability of a practical multicyclic sorption-based water harvester with improved water yield. Water Res 211:118029. https://doi.org/10.1016/j.watres.2021.118029

Xu W, Yaghi OM (2020) Metal-organic frameworks for water harvesting from air, anywhere, anytime. ACS Cent Sci 6:1348–1354. https://doi.org/10.1021/acscentsci.0c00678

Xu J, Li T, Chao J et al (2020) Efficient solar-driven water harvesting from arid air with metal-organic frameworks modified by hygroscopic salt. Angew Chemie - Int Ed 59:5202–5210. https://doi.org/10.1002/anie.201915170

Xu J, Li T, Yan T et al (2021) Ultrahigh solar-driven atmospheric water production enabled by scalable rapid-cycling water harvester with vertically aligned nanocomposite sorbent. Energy Environ Sci 14:5979–5994. https://doi.org/10.1039/d1ee01723c

Yang J, Zhang X, Qu H et al (2020) A moisture-hungry copper complex harvesting air moisture for potable water and autonomous urban agriculture. Adv Mater 32:1–8. https://doi.org/10.1002/adma.202002936

Yao H, Zhang P, Huang Y et al (2020) Highly efficient clean water production from contaminated air with a wide humidity range. Adv Mater 32:1–8. https://doi.org/10.1002/adma.201905875

Yuan Y, Zhang H, Yang F et al (2016) Inorganic composite sorbents for water vapor sorption: a research progress. Renew Sustain Energy Rev 54:761–776. https://doi.org/10.1016/j.rser.2015.10.069

Zhao F, Zhou X, Liu Y et al (2019) Super moisture-absorbent gels for all-weather atmospheric water harvesting. Adv Mater 31:1–7. https://doi.org/10.1002/adma.201806446

Zhao H, Lei M, Liu T et al (2020) Synthesis of composite material HKUST-1/LiCl with high water uptake for water extraction from atmospheric air. Inorganica Chim Acta 511:119842. https://doi.org/10.1016/j.ica.2020.119842

Zheng X, Wang R (2019) Microstructure and sorption performance of consolidated composites impregnated with LiCl. Int J Refrig 98:452–458. https://doi.org/10.1016/j.ijrefrig.2018.11.031

Zheng X, Wang LW, Wang RZ et al (2014) Thermal conductivity, pore structure and adsorption performance of compact composite silica gel. Int J Heat Mass Transf 68:435–443. https://doi.org/10.1016/j.ijheatmasstransfer.2013.09.075

Zhou X, Lu H, Zhao F, Yu G (2020) Atmospheric water harvesting: a review of material and structural designs. ACS Mater Lett 2:671–684. https://doi.org/10.1021/acsmaterialslett.0c00130

Chapter 6
Potential of Atmospheric Water Harvesting in Arid Regions: Case Studies

Bharti Budhalakoti, Sameer Kumar Maurya, Kanchna Bhatrola, N. C. Kothiyal, and Vaneet Kumar

Abbreviations

AWH	Atmospheric water harvesting
AWG	Atmospheric water generator
SWP	Specific water production
SEC	Specific energy consumption
RR	Recovery ratio
RH	Relative humidity
TEC	Thermoelectric cooling
VCC	Vapor compression cycle
PE	Polyethylene foil
MHI	Moisture harvesting index
SFC	Standard fog collectors
SC	Shading coefficient

B. Budhalakoti (✉) · S. K. Maurya · K. Bhatrola · N. C. Kothiyal
Department of Chemistry, Dr. B. R. Ambedkar National Institute of Technology Jalandhar, GT Road, Amritsar Bypass, Jalandhar, Punjab 144011, India
e-mail: bhartibudhalakoti@gmail.com

N. C. Kothiyal
e-mail: kothiyalnc@nitj.ac.in

V. Kumar
Department of Applied Sciences, CTIEMT, CT Group of Institutions, Jalandhar, Punjab, India

E. Fosso-Kankeu et al. (eds.), *Atmospheric Water Harvesting Development and Challenges*, Water Science and Technology Library 122,
https://doi.org/10.1007/978-3-031-21746-3_6

6.1 Introduction

Leonardo da Vinci once said "Water is the driving force of all nature". There is nothing more essential in life on earth than water. But in the present era of globalization and increased greenhouse gas emission, water scarcity has become a problem of concern. From sub-Saharan Africa to Asia and Central Australia, water is scarce. There is a struggle among people to access clean water either for drinking, cooking and cleaning purposes (Chartres and Varma 2010). Precisely, water scarcity attributes to lack of safe water supplies. With an increase in population and a global rise in surface temperature, access to fresh water dwindles. According to data reported by UNICEF, over two billion people live in countries where water supply is inadequate and over half of the world's population will be living in areas facing scarcity by 2025 (UNICEF 2020). Also mentioned in recent researches around 4.0 billion people face water scarcity at least one month a year and around half billion people face shortage yearly (Tu et al. 2018). The availability of water resources plays a very important role in improving the livelihood, household management and further economic development of the nation. Talking more precisely about arid regions, the word arid means dry that is faces a lack of precipitation. These regions receive less than 10 in. of rain per year which roughly amounts to 25 cm. Map 6.1 shows the arid region all around the world.

For instance, The Atacama Desert in Chile is the driest place on Earth (Quade et al 2008). On an average it receives 0.04 in. of precipitation every year. Here water

Map 6.1 Arid regions around the world

scarcity, drought like conditions and land degradation are some of the predominant features due to which human settlements are scattered around water resources like wells, springs and oases (Gaur and Squires 2018). Using seawater or purifying wastewater can be performed to makeup water scarcity by involving certain techniques like distillation, reverse osmosis or either filtration. But the only limitation being, these are inaccessible in landlocked regions. People residing in either arid or semi-arid regions are well aware regarding water that exists in atmosphere (Tu et al. 2018). Atmospheric water can be considered as an alternative which do not depend upon either hydrological or geographical conditions. The moisture present in atmosphere can act as a potential source which will be able to provide us with large amount of water and acts as a renewable source of energy (Tu et al. 2018). There are basically three types of atmospheric water namely: fog which is close to the ground, clouds and water vapor present in atmosphere (Beysens and Milimouk 2000). The availability of atmospheric water depends on the water holding capacity of the atmosphere which can hold up to 10% of the sources either in the form of vapor or in the form of droplets (Zhou et al. 2020). According to US geological survey, the estimate of water in atmosphere in given time is equals to 3100 cubic miles. Therefore, it can be considered that natural water cycle or hydrological cycle helps to maintain water supply. Atmospheric water harvesting (AWH) can act as a potent tool to rejuvenate fresh water and further help to overcome distant transportation of water facilities to remotest areas. The Technology incorporated for water harvesting must be based on the following criteria: cost efficiency, must be scalable and operating tendency either yearly or at least for monsoon season. The functionality of Atmospheric Water generator (AWG) can be as followed: firstly moisture is captured from the air and further the moisture which is captured is condensed to water. The two processes condensation as well as separation may consume energy which can be either wind or solar energy. The three parameters used to review the working of AWG can be given as: SWP or specific water production, SEC or specific energy consumption and RR or recovery ratio. A brief classification of AWG can be given as shown in Fig. 6.1 (Tu et al. 2018).

Significantly, AWH can also be defined as decentralized water production. Scientists and researchers have undergone certain innovations to make use of this technology. It comprises of mainly three categories which include dewing, fog-harvesting and sorption based technique. The process that enables small droplets to grow can be defined as fog harvesting. Dewing technology involves cooling air below its dew point and further collecting the condensate. Talking about applicability of these two methods they are best suited for areas with water shortage having high relative humidity (RH). However, talking about arid areas sorption based technique is feasible which require sorbent in order to adsorb water and further desorption takes place (Wang et al. 2019). Few others include TEC that is thermoelectric cooling, solar chimneys, VCC that is vapor compression cycle, using variety of adsorbents or membranes, active cooling condensation technology, radiative cooling system (Sultan et al. 2021). In this chapter we will be dealing with variety of methods, systems and case studies in order to have a closer look at atmospheric water harvesting.

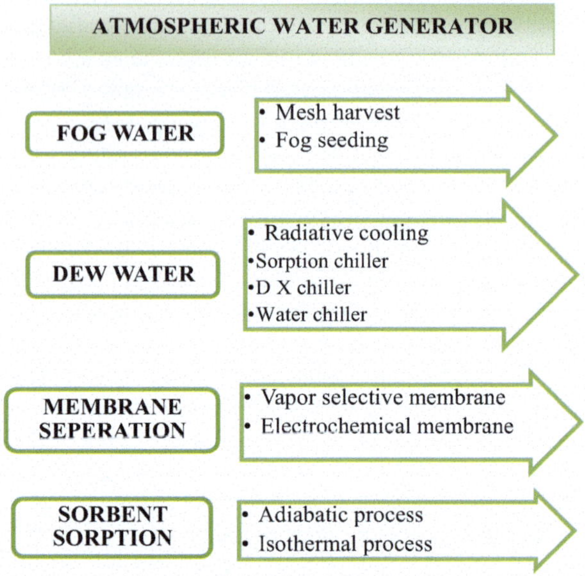

Fig. 6.1 A brief classification of AWG

6.2 Methods for Atmospheric Water Harvesting

Few of the methods that can be used for AWH includes.

6.2.1 Natural Harvesting (Tu et al. 2018)

Hales in the year 1727 carried out first study involving plants which are involved in absorbing moisture. Stone in the year 1957 carried out study which comprised of absorption of water and dew concentration in plants. The work was extended by Malik et al. who carried out a thorough study in the year 2014. His work was based on directing moisture in plants as well as animals like lizards, beetles etc.

6.2.2 Early Atmospheric Water Generator (Tu et al. 2018)

Considering dew as a source of water can be dated back to centuries. However, condensers first appeared in twentieth century. During sixth century BC it was Greeks who used dew condensers in order to meet their demand for water. A bowl shaped stone condenser was first built by F.I. Zibold during the period 1905–1912. It was considered similar to Greek condenser, still challenged by a few. However few of the researchers were inspired by Zibold's experiment and another type of condenser known as "aerial wells" was designed by Chaptal and Knapen (Chaptal 1930; Knapen 1928). These were designed so as to take advantage from temperature variation during day and night. Water was successfully condensed using these installations but amount produced was quite less.

6.2.3 *Modern Atmospheric Water Generators (Tu et al. 2018)*

Owing to availability of water different types of technologies can be classified as: fog harvesting (Fessehaye et al. 2014; Klemm et al. 2012), rain collection (DeFelice and Axisa 2017; Bruintjes 1999; Wang et al. 2016), ambient air cooling, vapor condensation and dew water collection (Dai et al. 2018).

6.2.4 *Rain Collection*

Artificial rain collection or weather modification or cloud seeding is a process by which precipitation may increase in the troposphere region due to presence of water abundant clouds. The process of cloud seeding is a traditional method of rainmaking which has been in use since 1940s. It involves an aircraft that injects AgI or silver iodide in atmosphere. These chemicals act as particles which provides surface for condensation which further leads to formation of water droplets. This operation was carried out throughout the dry season. A diagram depicting cloud seeding can be given in Fig. 6.2 as.

Example can be considered of Indonesia which in the year 2020 started cloud seeding so as to keep forest fires at bay. Advantages of rain collection or cloud seeding can be summed up in Fig. 6.3.

However, no evidence has yet been provided which proves that similar process could be attained at ground level.

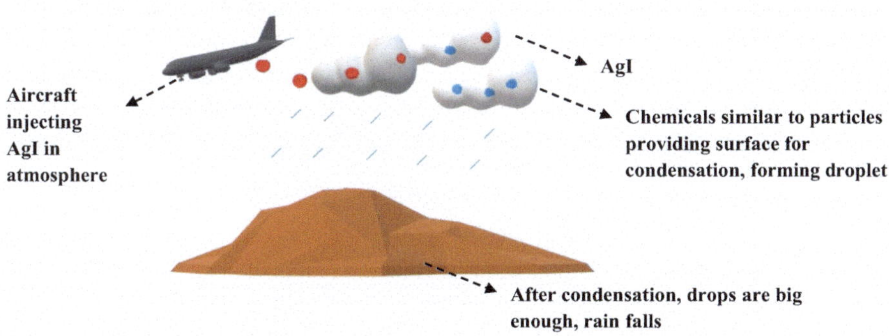

Fig. 6.2 Diagram depicting the process of cloud seeding

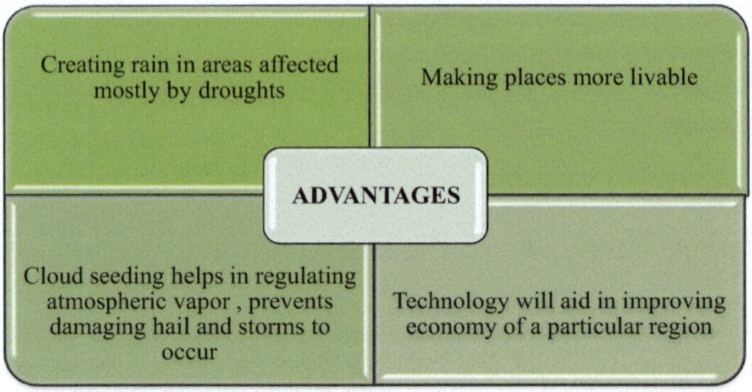

Fig. 6.3 Advantages of artificial rain collection

6.2.5 Fog Harvesting (Maleki et al. 2021)

The age old conventional method of fog harvesting has been performed in the region of Middle Eastern deserts for around 2000 years. The diameter of fog droplets lies between 1–50 μm (Ritter et al. 2008). It is obtained from water loss via process of evapotranspiration that gives rise to humid air either over sea or land. In case of radiation fog, it is obtained overnight because of condensation of vapor present above cooling ground (Straub et al. 2012). Fog can be considered as an important source of moisture all over the world (Dawson 1998). The coastal watersheds of California where redwood forest dominates, almost 34% of total annual water originates from fog which is dripped off from the redwood trees. This is a prime source for water collection during summers when precipitation drops off 25 mm/month (Domen et al. 2014). In Chile too the importance of fog-water is demonstrated by coastal rainforest present over mountain top. The amount of rainfall received in this region accounts up to 147 mm annually while the amount of contribution met up by fog water accounts to additional 200 mm. Talking about region where there is little or no rainfall, both flora and fauna survives on fog water. For instance, beetles of Namib Desert have adapted themselves in an area that receives annually less than 12 mm of rainfall. This is because a beetle back is made up of hydrophobic surface further covered by hydrophilic bump. It collects fog water which is drained from channels into beetle's mouth (Parker and Lawrence 2001). Later fog fences came into use. This technique can be mainly used in coastal areas where fog is brought in by inland winds. A mesh material kept perpendicular to wind collects fog droplets. During a foggy environment water droplets are trapped and due to coalescence when droplets grow these are collected in a water tank via gutter (Tu et al. 2018). The size of mesh depends upon quantity of water and available space and therefore can vary from 1 to 100 m (Maleki et al. 2021). In northern Chile, the first experiments with nets were conducted in the year 1956. Experiments based on above mentioned idea were conducted on a huge scale in the year 1987 by Schemenauer and colleagues covering some of the arid regions comprising Namibia in West Africa, Oman and Saudi Arabia

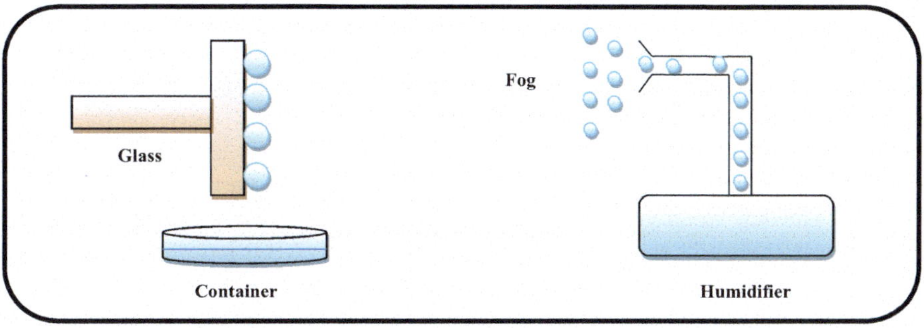

Fig. 6.4 Schematic representation of fog harvesting

Table 6.1 Data for AWH in three different cities

Station	Amount of water harvested
Mombasa	1.8–3.7 L/m^2 per day
Kisumu	0.8–2.2 L/m^2 per day
Nairobi	1–2.5 L/m^2 per day

in Middle East and Peru in South America. In the above mentioned projects it was observed that collectors were far away from the residents which required pipelines to deliver water. This increased the infrastructure cost which was quite uneconomical. A schematic diagram for fog harvesting can be shown as given in Fig. 6.4 (Zhang et al. 2020). It was observed by Abdul Wahab et al. the future and prospective of residential type collectors built close to the houses (Tu et al. 2018).

If we consider AWH in case of Kenya at 10% air humidity following data was obtained for three different cities (Ngaina et al. 2014) given in Table 6.1.

In case of Nepal, the amount of water daily produced equals to 500 L and in dry season half of its quantity is produced. South American states of Ecuador, Chile and Peru are dependent on mesh technique to make use of atmospheric water for irrigation purposes. East African state of Eritrea produced 12,000 L of water per day using mesh size of 1600 m. According to IUCN i.e. International Development Research Centre, other countries that make use of this technique are China, Cape Verde, South Africa, Sri Lanka and Yemen. Research and innovation are still going on to provide good quality of meshes which will further enhance production of water.

6.2.6 Dew Water Collection (Tu et al. 2018)

Dew water collection precisely known as a non-conventional source that helps to makeup water shortage in certain arid/semiarid regions. Therefore it can be regarded as an alternative source for water supply where dew formation is favored; however these systems are quite rare which proposes this technique to be under-explored.

These collectors are divided into two important classes comprising active dew condenser and passive or radiative condenser. A schematic diagram for passive dew collection can be shown as given in Fig. 6.5 (Zamir et al. 2019).

Research work was being carried out somewhere around early 1960s targetting passive condensers (Gindel 1965). It mainly focused on material being used in condenser, designing and climatic features that affects the amount of dew water collected. The dew water yield recorded in arid regions lie between 0.3–0.6 L/day/m^2 of surface area (Muselli et al. 2009; Maestre-Valero et al. 2011; Lekouch et al. 2012). This technique is quite popular when compared to cloud seeding and fog water collection because it is cost efficient and is hardly affected by geographical and climatic parameters (Khalil et al. 2016; Sharan et al. 2017; Girja 2008). Several factors are involved for dew formation that includes humidity, temperature variation and vapor pressure. These condensers work normally during nighttime at an ambient temperature time when compared to higher temperature during daytime due to solar radiation. The dew collection process has been standardized by International Organization for Dew Utilization on the basis of their instrumentation and type of methodology incorporated. PE or polyethylene foil has been recommended by the organization in order to process standard material. During 1930s active dew condensers were introduced in market but wide research and innovation took place once mechanical refrigerators were commercialized during 1980s. These active condensers designed to deal with water quantity as well as quality issue proves to be an innovative option and are quite

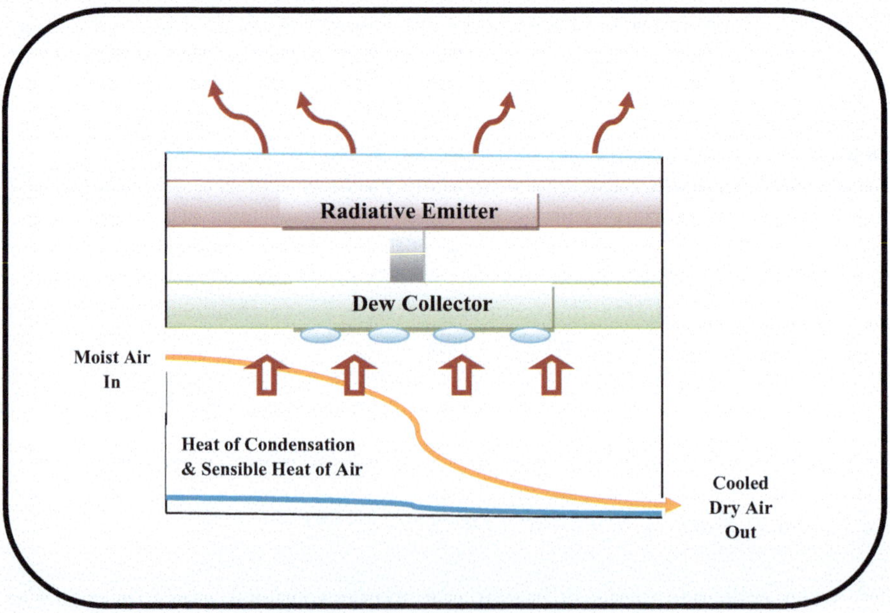

Fig. 6.5 Schematic diagram for passive dew collection

similar to a dehumidifier. Use of sorption chillers for harvesting can work in day as well as night time (in presence of heat storage).

6.2.7 *Condensation of Vapor*

The process of condensation can be explained as change of vapor into liquid state. It can also be stated as reverse of evaporation. Extraction of atmospheric water can be performed using vapor condensation that includes cooling of moist air below dew point followed by a cold surface contact. However, this process limits condensation due to latent heat release amounting to 2500 $\frac{kJ}{kgw}$ from the surface keeping temperature below its dew point (Maleki et al. 2021). Study performed in Namib Desert (Southern African region), which is supposed to be arid for at least 55 million years receives rainfall < 10 mm approximately 0.39 in. and is almost barren. Darkling beetles found in this desert use radiative vapor condensation technique in order to collect water. Dew is generated by the help of their bodies which acts like a cooling surface that radiates thermal energy towards night sky (Guadarrama Cetina et al. 2014). Similar technique is observed in case of commercial dew condensers (Nilsson et al. 1994) where radiative cooling helps to extract atmospheric moisture by forming dew when directed towards the night sky (Muselli et al. 2006). This process provides an expected yield of 0.8 $\frac{L}{m^2 d}$ but actual results provided a low water yield (Maleki et al. 2021). It was observed by Fan et al. that during daytime radiative cooling can be observed at subambient temperatures by using a highly efficient reflector along with thermal emitter in mid infrared (mid-IR) range (Nilsson et al. 1994). Talking about the process of condensation it can be performed in two ways: firstly, cooling of air to its dew point and secondly, saturation with vapor upto maximum limit that ultimately leads to condensation. A detailed discussion regarding these phenomena can be given as.

6.2.7.1 Dew Point

It can be defined as the temperature where condensation takes place. Talking about air temperature they either reaches the dew point or falls below it, for instance in case of night time. This is the reason behind we often visualize dew coated over lawns, houses etc. When observed in arid regions, the frequency of dew in comparison to rain is higher and for much longer time. The Negev desert region consists of brown and dusty mountains lies in southern Israel. It receives very little rain because it is situated east to Sahara. The frequency of dew events can be summed up around 55% during dew days and the duration was around 7.6 h/night annually (Zangvil 1996). In Almeria, a city located in Southeast Spain the frequency was accounted to around 50% and duration was about 6.5 h/day (Moro et al. 2007). A study conducted in Shaanxi province of China which was a three year field provided some important information (Jia et al. 2019).

1. Dew is classified as an important resource majorly in arid or semi-arid regions owing to its larger frequency. The total number of days when dew is observed that is dew days accounts to 2.6 times to that of days when rainfall is observed. Also, when comparing the dew amount it was around on an average 32.8 mm or 1/18th of total amount of rainfall.
2. The maximum amount of dew which was observed daily was about 0.88 mm.
3. A strong correlation was observed between dew amount and relative humidity, and a weak correlation with wind direction, its speed, amount of moisture and temperature of soil.

6.2.7.2 Saturation

While defining the term saturation let us consider a simple example of clouds. Clouds are defined as collection of water droplets. As the amount of water vapor increases, they are saturated. These saturated clouds have little tendency to hold back water. Thus condensation of vapor takes place which ultimately leads to rainfall.

6.2.8 Cooling of Ambient Air

Another technique that can be approached in order to harvest atmospheric water is using cooling of ambient air, which is performed using an electric compression expansion device (Wahlgren 2001, 2014). However, the process majorly depends on meteorological conditions. These systems work more efficiently when water requirement is small. MHI or Moisture Harvesting Index helps to assess the appropriateness of climate for moisture harvesting via direct electric cooling. Here, the temperature difference generated due to congregation of ambient air on a cold surface results in transfer of heat from air to surface. This results in decrease in the temperature of the air that is near to the surface and further vapor condensation which surpass moisture saturation capacity of chilled air. MHI signifies ratio of energy incorporated for condensation of water to total energy required for cooling both condensable and incondensable gasses (Gido et al. 2016). Ouagadougou's capital city of Burkina Faso is classified as hot semi-arid region under the Köppen-Geiger classification and has an annual rainfall of around 800 mm or 31 in. This region is expected to have suitable moisture harvesting in summers than in winters and an average MHI of 0.32. Cabanatuan, Philippines with a tropical wet and dry climate possesses an average MHI of 0.59. Table 6.2 depicts variations in AMH calculated using direct cooling for varied locations.

It is observed that among all studied locations (Table 6.2) for AMH the most appropriate locations is Cabanatuan, Philippines having an average MHI value of 0.59 over a period of ten years. However, several improvements and innovation in last 20 years have been performed for improving efficiency and productivity with

Table 6.2 Variations in AMH for different locations

Location		Annual rainfall (mm)	Distance to sea (km)	Altitude (m)	Average MHI
Country	City				
Burkina Faso	Ouagadougou	792	800	307	0.32
India	Rajkot	690	100	122	0.41
	Panjim	2750	2	59	0.55
Philippines	Cabanatuan	2151	60	32	0.59
Syria	Damascus	204	80	615	0.16
Kenya	Mombasa	1050	1	61	0.55
	Nairobi	1050	250	1623	0.45
Yemen	Aden	30	1	3	0.5
	Sanaa	200	150	2206	0.1

respect to atmospheric conditions, different collector designs and material being used.

6.3 Atmospheric Water Harvesting in Arid and Semi Arid Regions (Case Studies)

In this section we will be studying few regions where different techniques were incorporated in order to harvest atmospheric water.

6.3.1 Namibia-Fog as a Source (Shanyengana et al. 2002)

Namibia, a Southern African country is a hyperarid country where it is difficult to find freshwater. Map 6.2 shows the location of Namibia. The fog is observed for about 60–200 days in a year. The precipitation obtained using fog is somehow greater as compared to rainfall in this area and could be considered much as a reliable source for flora, fauna and human settlement. However, for human settlements fog collection started in the year 1995. In the coastal region of Swakopmund, an average annual rainfall of 18 mm while in the inland regions of Gobabeb an average of 21 mm is observed (Lancaster et al. 1984; Nagel 1959). Also, it is quite common to encounter no rainfall for consecutive years. An example can be considered of Swakopmund where no rainfall was recorded for a period of 10 years. In order to measure fog deposition a wire mesh cylindrical in shape was brought to use which was kept over a rain gauge (Lancaster et al. 1984; Henschel et al. 1998). This experiment was performed at four different sites which include Gobabeb, Vogelfederberg, Ganab and Kleinberg. Talking about inland areas it was observed that fog deposition was

Map 6.2 Namibia, Southern African country

Table 6.3 Results for fog collection in Namibia

Location	Average fog collection (ml/m^2)
Swartbank	2390/wet day
Gobabeb	508/wet day
Klipneus	3308/wet day

around seven times when compared to rainfall (Lancaster et al. 1984; Henschel et al. 1998; Nagel 1959; Nagel 1962; Pietruszka and Seely 1985). SFC or Standard Fog Collectors are used for investigating fog collection made up of a polypropylene mesh (Schemenauer and Cereceda 1994). Results obtained for fog collection can be summed up in Table 6.3 (Henschel et al. 1998) as.

It was observed that inspite of poor quantity of available traditional resources the problem of water scarcity can be met by fog collection. This prime source acts as a viable resource in rural as well as urban settlements.

6.3.2 Chile-Dew Water Harvesting

The Republic of Chile lies in western part of South America. The northern part comprises of desert type climate for instance Atacama Desert which is the hottest desert in the world. Infact some of the regions of the Desert have never received rain at all. A large range of temperature can be observed daily which is equivalent to 30 °C. Moving towards Central Chile; the climate present is Mediterranean where summers are long and hot while winters are wet and cool. A rainy climate is observed when

Map 6.3 Chile, Western part of South America

moving towards southern part. Early mornings in Santiago observed photochemical smog. Map 6.3 shows the location.

In order to determine nitrite effect, rain and dew in the city was analyzed by Rubio et al. (2002). On 1 m^2 PTFE surface, 0.127 Lm^{-2} d^{-1} dew was reported (Carvajal et al. 2018). The first fog collectors of Chile were developed in the year 1987 in El Tofo by Schemenauer. This further led to projects related with fog collection in Chungungo. A number of collections projects till then have been established in South Africa, Israel, Peru, Cape Verde Islands, Nepal etc. The project carried in Combarbalá was integrated in a household comprised of 36 m^2 galvanized steel roof at 15°. A paint comprising of aluminosilicate minerals which is highly emissive in nature is coated on the rooftop so that the yield of dew can increase. The city of Combarbalá was considered due to following reasons: Firstly, it signifies the temperate steppe climate which covers around 20,000 km^2. Secondly, this area is suffering from desertification.

6.3.3 Rajasthan-Water Harvesting and Moisture Conservation (Narain et al. 2005)

The largest state of India, Rajasthan covers 10.5% of the country's area but the percentage share of its water resources is quite low accounting upto 1.15%. The

Map 6.4 Rajasthan, state in northern India

state is predominantly either arid or semi-arid where agriculture is the prime source of livelihood. Map 6.4 shows the location of Rajasthan. The water availability was estimated around 840 m^3 in the year 2001 and is expected to go below by the year 2050 somewhat around 439 m^3.

6.3.3.1 Water Harvesting and Its Potential

In order to increase availability of water especially during drought period water harvesting and conservation are taken in consideration. It involves rainfall concentration from a catchment further into a targeted area. For instance, the technique which is used to recharge groundwater is Khadin which is mostly practiced in hyper-arid portions and involves collecting runoff from rocky surface in an earthen embankment. Few other technologies involve moisture conservation which is cost effective and proves to be quite efficient to tackle drought like situation as well as increasing land productivity. Contour Bunding is one such technique in which water is conserved for rain-fed farming. It is found to be more suitable for soil bund which varies from 0.3 to 0.6 m in height. CVB or Contour vegetative barriers comprise grass species that are cost effective and environmental friendly which acts efficiently for in-situ soil conservation. Some of the native, locally available systems involve *Cenchrus ciliaris*, *Cenchrus munja* and *Cymbopogon jwarancusa. It* can therefore be concluded that the region prominently depends upon traditional systems for water conservation and harvesting including jhalara, tanka, bawari etc. However, more focus has to be put upon moisture conservation by contour bunding and practices.

Map 6.5 Syria, country in Western Asia

6.3.4 Syria-Atmospheric Water Harvesting Via Fog and Dew Water

Syria a country in Western Asia has a hot, dry and desert type of climate. The rate of precipitation has observed a decrease for about half a century in accordance with the current climatic conditions (Khalil et al. 2022; Mawed and Alshihabi 2014). An increase in drought pattern was also observed in rainy season. An increase in water availability is required for growing population as well as rapid industrial growth. The changing climatic conditions will affect the water resources in such a way that the ground water will observe a decrease by 1300 million cubic meters.

Map 6.5 shows the location of Syria. In these conditions, fog and dew water are an important resource. This could help in those regions where availability to fresh water is scarce for example East Syria. There were basically two models that were used for water collection which can be given as.

MODEL 1	MODEL 2
• Collecting area of 1.00 m^2. • Base is equal to 2 m that is above the ground. • Top is around 3 meters above the ground. • Use of polyprropylene mesh.	• Cylidrical fog collector • Comprising two bases; one at the top and other at the bottom. • Mesh placed on stainless steel, bit concave.

Table 6.4 Different types of mesh used in experiment

Mesh type	Shading coefficient (SC) %
Polypropylene single mesh	35
	50
Double polypropylene mesh	35
	50
Polyethylene single	60
Double polyethylene	60
Metal mesh	40

The environmental and climatic conditions play a major role in deciding the water collection. Some of the crucial factors comprises of temperature, direction of wind, relative as well as absolute humidity, rainfall and dew point temperature (Olivier 2004). Different types of mesh were also used while performing the experiments. These are listed in Table 6.4.

Afterwards, the meshes are applied with a mixture of methylphenyl silicon and hydrophobic silica particles. It was observed that better results were obtained using double polypropylene mesh having a shading coefficient equals to 35%. Rest of the double mesh having a lager SC% gave lesser yields. The reason that was reported regarding this study was higher the ratio, smaller the openings. This prevents the water droplets to pass. That is why coating is performed so that water droplets can move downwards and won't stay longer in mesh openings.

6.4 Conclusion

There exists a huge possibility for the existence of plenty amount of water availability in atmosphere. Even in arid or semi-arid regions where amount of rainfall is either small or negligible, airborne moisture can act as an important source of water. In the last 20 years, the technology of atmospheric water harvesting as observed a significant boom so as to renewably make up water scarcity. In this chapter we came across different methods for harvesting. AWGs are devices that condenses vapor from the air. The two techniques used by these devices are cooling condensation and wet desiccation. The market for AWG is a growing market for rising water demand and growing water scarcity. Due to their capability in providing water for different purposes their requirement is increasing. Efforts are being put by companies to improve this growing technology. Several R&D projects and innovative ideas are being invested. Fog actually is a natural phenomenon in which tress for example redwoods in American Pacific Coast or the forest present in Peru and Chile, collect the water that is present in atmosphere to protect them in water scarcity. There efficiency can be attributed to two important question; how tall are they and what kind of structure does the leaves have. Few other factors that influence the water intensity are speed of wind, size of water droplets and its distribution. Technology based method for fog collection comprises of mesh facing the wind, so that the droplets gets trapped as air passes through it. The next one in line is dew water harvesting. It has significant importance since this technique when compared to fog water collection is not much affected by either geographically or climatically. Starting from early 1950, the significance of dew formation was observed by both hydrologists as well as physicists. Technology based active dew condensers or radiative dew condensers are involved for collection process. The advantages that atmospheric water harvesting holds are diverseness of supply, generating water supply off grid in arid, semi-arid and isolated regions. It is observed that water obtained by this technology provides safe water because it is completely isolated from either the surface water or the contaminated ground. Out of 17 SDGs, SDG-6 aims to provide universal access to clean and safe water. It has been regarded as an important aspect covering health, socioeconomic development and equity among all. AWH proves to be important for unexplored sources of water which is quite abundant and is accounted upto 3 trillion litres. It is a major chunk of water supply since it has been observed that there exists a seasonal fluctuation and majority of people have to face water scarcity. This technique can prove to be cost efficient and financially stable for low income countries where water access is affected by climate changes.

References

Beysens D, Milimouk I (2000) The case for alternative fresh water sources. Pour Les Res Altern En Eau, Secheresse 11(4):1–16

Bruintjes RT (1999) A review of cloud seeding experiments to enhance precipitation and some new prospects. Bull Am Meteor Soc 80(5):805–820

Carvajal D, Minonzio JG, Casanga E, Muñoz J, Aracena A, Montecinos S, Beysens D et al (2018) Roof-integrated dew water harvesting in Combarbalá, Chile. J Water Supply Res Technol—AQUA 67(4):357–374

Chaptal L (1930) La lutte contre la sécheresse par la captation de la vapeur d'eau atmosphérique. Journal D'agriculture Traditionnelle Et De Botanique Appliquée 10(107):566–568

Chartres C, Varma S (2010) Out of water: from abundance to scarcity and how to solve the worlds water problems. FT Press, Upper Saddle river, New Jersey

Dai X, Sun N, Nielsen SO, Stogin BB, Wang J, Yang S, Wong TS et al (2018) Hydrophilic directional slippery rough surfaces for water harvesting. Sci Adv 4(3):eaaq0919

Dawson TE (1998) Fog in the California redwood forest: ecosystem inputs and use by plants. Oecologia 117(4):476–485

DeFelice TP, Axisa D (2017) Modern and prospective technologies for weather modification activities: developing a framework for integrating autonomous unmanned aircraft systems. Atmos Res 193:173–183

Domen JK, Stringfellow WT, Camarillo MK, Gulati S et al (2014) Fog water as an alternative and sustainable water resource. Clean Technol Environ Policy 16(2):235–249

Fessehaye M, Abdul Wahab SA, Savage MJ, Kohler T, Gherezghiher T, Hurni H et al (2014) Fog-water collection for community use. Renew Sustain Energy Rev 29:52–62

Gaur MK, Squires VR (2018) Geographic extent and characteristics of the world's arid zones and their peoples. Climate variability impacts on land use and livelihoods in drylands. Springer, Cham, pp 3–20

Gido B, Friedler E, Broday DM et al (2016) Assessment of atmospheric moisture harvesting by direct cooling. Atmos Res 182:156–162

Gindel I (1965) Irrigation of plants with atmospheric water within the desert. Nature 207(5002):1173–1175

Girja S (2008) Harvesting dew water using radiative-cooled condenser to supplement drinking water supply in hot arid coastal area of north-west India. Paper presented at the Agricultural and biosystems engineering for a sustainable world, International conference on agricultural engineering, Hersonissos, Crete, Greece, European Society of Agricultural Engineers, 23–25 June 2008

Guadarrama Cetina J, Mongruel A, Medici MG, Baquero E, Parker AR, Milimouk Melnytchuk I, Beysens D et al (2014) Dew condensation on desert beetle skin. Eur Phys J E 37(11):1–6

Henschel JR, Mtuleni V, Gruntkowski N, Seely MK, Shayengana E et al (1998) Namfog: Namibian applications of fog-collecting systems: phase 1. Evaluation of fog-water harvesting. Occasional paper No 8, DRFN

Jia Z, Wang Z, Wang H et al (2019) Characteristics of dew formation in the semi-arid Loess Plateau of central Shaanxi Province, China. Water 11(1):126

Khalil B, Adamowski J, Shabbir A, Jang C, Rojas M, Reilly K, Ozga Zielinski B et al (2016) A review: dew water collection from radiative passive collectors to recent developments of active collectors. Sustain Water Res Manag 2(1):71–86

Khalil MM, Kara Ali A, Assad M et al (2022) Potential of harvesting water from fog and dew water over semi-arid and arid regions in Syria. Water Supply 22(1):874–882

Klemm O, Schemenauer RS, Lummerich A, Cereceda P, Marzol V, Corell D, Fessehaye GM et al (2012) Fog as a fresh-water resource: overview and perspectives. Ambio 41(3):221–234

Knapen A (1928) Memoires sur le puits aerien. Bulletine Society Ingenieuns Civils, 139–140

Lancaster J, Seely LN, MK, et al (1984) Climate of the central Namib Desert. Modoqua 1984(1):5–61

Lekouch I, Lekouch K, Muselli M, Mongruel A, Kabbachi B, Beysens D et al (2012) Rooftop dew, fog and rain collection in southwest Morocco and predictive dew modeling using neural networks. J Hydrol 448:60–72

Maestre Valero JF, Martinez Alvarez V, Baille A, Martín Górriz B, Gallego Elvira B et al (2011) Comparative analysis of two polyethylene foil materials for dew harvesting in a semi-arid climate. J Hydrol 410(1–2):84–91

Maleki M, Eslamian S, Hamouda B et al (2021) Principles and applications of atmospheric water harvesting. In: Handbook of water harvesting and conservation: basic concepts and fundamentals, pp 243–259

Mawed K, Alshihabi O (2014) Assessment of precipitation in Syria, trend analysis, during the period of (1955–2006). Arab J Arid Environ 7:50–58

Moro MJ, Were A, Villagarcía L, Cantón Y, Domingo F et al (2007) Dew measurement by Eddy covariance and wetness sensor in a semiarid ecosystem of SE Spain. J Hydrol 335(3–4):295–302

Muselli M, Beysens D, Milimouk I et al (2006) A comparative study of two large radiative dew water condensers. J Arid Environ 64(1):54–76

Muselli M, Beysens D, Mileta M, Milimouk I et al (2009) Dew and rain water collection in the Dalmatian Coast, Croatia. Atmos Res 92(4):455–463

Nagel JF (1959) Fog precipitation at Swakopmund. Weather Bureau Newslett 125:1–9

Nagel JF (1962) Fog precipitation measurements on Africa's southwest coast. Notos 11:51–60

Narain P, Khan MA, Singh, G et al (2005) Potential for water conservation and harvesting against drought in Rajasthan, India, Working paper 104

Ngaina JN, Muthama JN, Opere AO, Ininda JM., Ng'etich CK, Ongoma V, Mutai BK et al (2014) Potential of harvesting atmospheric water over urban cities in Kenya

Nilsson TMJ, Vargas WE, Niklasson GA, Granqvist CG et al (1994) Condensation of water by radiative cooling. Renew Energy 5(1–4):310–317

Olivier J (2004) Fog harvesting: An alternative source of water supply on the West Coast of South Africa. GeoJ 61(2):203–214

Parker AR, Lawrence CR (2001) Water capture by a desert beetle. Nature 414(6859):33–34
Pietruszka RD, Seely MK (1985) Predictability of two moisture sources in the Namib Desert. South African J Sci
Quade J, Rech JA, Betancourt JL, Latorre C, Quade B, Rylander KA, Fisher T et al (2008) Paleowetlands and regional climate change in the central Atacama Desert, Northern Chile. Quatern Res 69(3):343–360
Ritter A, Regalado CM, Aschan G et al (2008) Fog water collection in a subtropical elfin laurel forest of the Garajonay National Park (Canary Islands): a combined approach using artificial fog catchers and a physically based impaction model. J Hydrometeorol 9(5):920–935
Rubio MA, Lissi E, Villena G (2002) Nitrite in rain and dew in Santiago city, Chile. Its possible impact on the early morning start of the photochemical smog. Atmos Environ 36(2):293–297
Schemenauer RS, Cereceda P (1994) Fog collection's role in water planning for developing countries. In: Natural resources forum, vol 18, No 2. Blackwell Publishing Ltd, Oxford, UK, pp 91–100
Shanyengana ES, Henschel JR, Seely MK, Sanderson RD et al (2002) Exploring fog as a supplementary water source in Namibia. Atmos Res 64(1–4):251–259
Sharan G, Roy AK, Royon L, Mongruel A, Beysens D et al (2017) Dew plant for bottling water. J Clean Prod 155:83–92
Straub DJ, Hutchings JW, Herckes P et al (2012) Measurements of fog composition at a rural site. Atmos Environ 47:195–205
Sultan M, Bilal M, Miyazaki T, Sajjad U, Ahmad F et al (2021) Adsorption-based atmospheric water harvesting. In: Technology fundamentals and energy-efficient adsorbents. Technology in agriculture, p 369
Tu Y, Wang R, Zhang Y, Wang J et al (2018) Progress and expectation of atmospheric water harvesting. Joule 2(8):1452–1475
UNICEF (2020) Water scarcity, addressing the growing lack of available water to meet children's needs. http://www.unicef.org. Accessed 1 March 2020
Wahlgren RV (2001) Atmospheric water vapour processor designs for potable water production: a review. Water Res 35(1):1–22
Wahlgren RV (2014) Another water resource for Caribbean countries: water-from-air. Paper presented at the Caribbean Water & Wastewater Association, twenty-third annual water & wastewater conference and exhibition, Atlantis, Paradise Island, Bahamas, 6–11 Oct 2014
Wang G, Zhong D, Li T, Wei J, Huang Y, Fu X, Zhang Y et al (2016) Sky River: Discovery, concept, and implications for future research. Sci Sinica Technol 46(6):649–656
Wang J, Dang Y, Meguerdichian AG, Dissanayake S, Kankanam Kapuge T, Bamonte S, Suib SL et al (2019) Water harvesting from the atmosphere in arid areas with manganese dioxide. Environ Sci Technol Lett 7(1):48–53
Zamir Y, Drechsler N, Howell J et al (2019) Deep radiative cooling passive dew collection. arXiv preprint arXiv:1909.10937
Zangvil A (1996) Six years of dew observations in the Negev Desert, Israel. J Arid Environ 32(4):361–371
Zhang J, Chen F, Lu Y, Zhang Z, Liu J, Chen Y, Parkin IP et al (2020) Superhydrophilic–superhydrophobic patterned surfaces on glass substrate for water harvesting. J Mater Sci 55(2):498–508
Zhou X, Lu H, Zhao F, Yu G et al (2020) Atmospheric water harvesting: a review of material and structural designs. ACS Mater Lett 2(7):671–684

Chapter 7
Sustainability of Atmospheric Water Harvesting in the Remote Areas

Rajeev Jindal, Vasudha Vaid, Khushbu, Kuljit Kaur, Priti Wadhera, and Rachna Sharma

7.1 Introduction

Globally, the numbers of people who are living in regions of physical water scarcity are 1.2 billion. 2.1 billion people lack access to water, while 4.5 billion people have insufficient sanitation, and safe and drinkable water (Sleiti et al. 2021). One-third of the world's large aquifer bodies are in distress. This leads to the risk of spread of water-borne diseases like cholera and typhoid fever which leads to the demise of 340,000 children under the age of five each year owing to the diarrheal diseases alone. Several large cities across the globe are at risk of a water crisis and it is imagined to be enhanced in most the countries in coming decades. Qatar, Israel, and Lebanon are ranked first, second and third in the list of countries dealing with the worst water stress, and many more countries that are currently suffering from water crises such as Iran, Jordan, Libya, and Kuwait. The water scarcity issue is not only affecting human beings but directly or indirectly, it is affecting all the ecosystems and all the natural pathways that need water. With an increase in population, the demands for resources also get enhanced resulting in an extra burden on freshwater resources which ultimately depletes the water aquifers and surface water in many places leading to the water crisis. Water crisis can further result in prolonged droughts, forced migration, and other emergencies in the same region (Klemm et al. 2012a, b). Therefore, water

R. Jindal (✉) · V. Vaid · Khushbu · R. Sharma
Polymer and Nanomaterial Lab, Department of Chemistry, Dr. B R Ambedkar National Institute of Technology, Jalandhar, Punjab 144011, India
e-mail: jindalr@nitj.ac.in

K. Kaur
Faculty of Natural Science, GNA University, Phagwara, Punjab 144401, India

P. Wadhera
Department of Applied Chemistry, Humanities and Management, Sardar Beant Singh State University, Gurdaspur, Punjab 143530, India

E. Fosso-Kankeu et al. (eds.), *Atmospheric Water Harvesting Development and Challenges*, Water Science and Technology Library 122,
https://doi.org/10.1007/978-3-031-21746-3_7

scarcity is a growing concern worldwide that needs urgent and quick action. Hence, the near future demands finding alternate water resources and the development of advanced technology to produce drinkable water. Several methods are being utilized for water harvesting globally namely desalination of water, rainwater harvesting and storage, and groundwater harvesting (Klemm et al. 2012a; Malik et al. 2014; Jarimi et al. 2020; LaPotin et al. 2019; Olivier and De 2002; Alqadami et al. 2017; Fessehaye et al. 2014; Taylor et al. 2013; Schemenauer 2014; Macedonio et al. 2012; Fath et al. 2011; Kogan and Trahtman 2003; Vuollekoski et al. 2015; Gido et al. 2016; Ejeian et al. 2020). One of the well-developed methods is the conversion of seawater into drinkable form (desalination of seawater) by using Reverse Osmosis membrane technology. Reverse Osmosis is a highly cost-effective technique and is not practiced widely. Secondly, desalination of seawater cannot be used in areas that have no access to the seawater. Moreover, transportation costs from one region to another will need massive investments in infrastructure and energy requirements. For these techniques to work, water must be present in the liquid form to harvest but when no such availability is there, harvesting of atmospheric water becomes crucial. Rainwater harvesting is a cheap and easy approach that requires little expertise and has many advantages in remote areas. But atmospheric water harvesting requires special expertise and is important in the area where the availability of water in liquid form is not possible to harvest. Therefore, atmospheric water harvesting is now attaining more attention from researchers worldwide. The water present in the form of vapors in the atmosphere of our planet is around six times the water present in the rivers at any given time. For a couple of decades, atmospheric water harvesting techniques are commercially available and around seventy companies are selling the equipment for the same purpose. The equipment is used in remote areas like drought-affected regions, where the drinking water is contaminated, during any disaster, and by the military during operations.

In atmospheric water harvesting, water can be harvested either from fog or vapors. Fog is tiny water droplets suspended in the air at or near the earth's surface and most importantly it is visible to the naked eye. The ways which can be used for harvesting fog can further be classified as traditional and modern methods (Klemm et al. 2012a, b; Batisha 2015; Park et al. 2013). Atmospheric water harvesting from fog is quite promising and cheap harvesting technology to generate drinking water, irrigation, and forest refurbishment in dry areas. Secondly, the water can be harvested from the vapors, which are not visible to the naked eyes and are formed by the evaporation of water and sublimation of ice. When the water vapors are condensed at a surface cooled temperature or the warm vapors come in contact with the cool surface that is the temperature below the dew point temperature, the water vapor gets condensed to the 'dew water'. A large number of adsorption materials like silica gel, zeolites, and metal–organic frameworks have been utilized in atmospheric water harvesting. These structures are capable of trapping water molecules in their structures and tend to harvest atmospheric water without any toxic emissions (Sleiti et al. 2021). Therefore, this topic of research has been quite interesting and needs substantial exploration nowadays. In this chapter, we will discuss the diverse atmospheric water harvesting technologies of fog as well vapors in detail, their working and utilization.

The ultimate goal of developing an atmospheric water harvester is to create a harvester that can generate water from its vapor state regardless of humidity level, geographical location, cheapness, and can utilize the local materials. We hope this chapter will inspire the researchers to think out of the box and to develop their own, unique and innovative ideas to create a sustainable atmospheric water harvester that can be utilized in the future for mankind to decrease water scarcity, enhance the economic development in remote areas (Sleiti et al. 2021; Jarimi et al. 2020; Munoz-garcia and Moreda 2005; Bui et al. 2017a; Tu et al. 2018a) where the annual rainfall is negligible, maintain the water table and to fulfill the needs of all living beings.

7.2 Techniques of Atmospheric Water Harvesting

Atmospheric water harvesting (AWH) has undergone evidential advancement in the last 25 years, as it is a promising way to fulfill the demand for fresh water and thus helps to sustain life even in extreme weather conditions. It is mainly categorized into fog water collection (FWC) (Klemm et al. 2012a; Kogan and Trahtman 2003) and dew water collection (DWC) (Tomaszkiewicz et al. 2015).

7.2.1 Fog Water Collection

7.2.1.1 Fog

Fog is a collection of suspended water droplets in air or near the surface of the earth (Klemm et al. 2012a; Abdul-Wahab and Lea 2008). It reduces the visibility within a distance of 1 km and this feature helps to distinguish it from haze and mist (Kogan and Trahtman 2003). The size of the droplets ranges from 1 to 40 μm in diameter, with the ability of larger drops to be efficiently collected by narrow fibers (Schemenauer and Cereceda 1992, 1994a). Since the falling fog droplets have very low velocity, so they move almost horizontally and can be collected on a vertical frame.

7.2.1.2 Fog Collection

FWC is a low-cost, low-maintenance, and sustainable method in areas that receive scanty rainfall and where fog formation is a regular episode (Domen et al. 2014). It utilizes simple technology to extract moisture from the fog by using vertical structures consisting of mesh. The mesh captures water droplets on exposure to fog carried by the wind. These water droplets coalesce to form bigger droplet which runs down to the attached gutter and is finally collected into storage tanks. For an efficient fog extraction program, the following criteria should be met as stated by Schemenauer and Cereceda (1994a, b):

1. Frequent persistent episodes of fog;
2. High-moisture content in fog;
3. Persistent winds.

Different types of fog collectors result in varying amounts of fog collection depending upon the size, shape, and material of mesh utilized, the surface area of the collector, and wind speed. Fog collectors are categorised as Standard fog collectors (SFCs) and Large fog collectors (LFCs) (Klemm et al. 2012a; Taylor et al. 2013). SFCs are used for exploratory studies on a small-scale level to estimate the quantity of water collected from fog under specific conditions at a particular site. The detail regarding the construction and use of SFCs is given by Schemenauer and Cereceda (Alqadami et al. 2017). It has (1×1) m^2 flat surface installed perpendicularly to the direction of prevailing winds and is installed 2 m above the ground from its base for efficient fog collection. Several investigatory experiments were done by Schemenauer and his colleague using SFC and the results inspired others to initiate and implement similar projects in many parts of the world. LFCs are used for actual fog collection and are similar to SFCs, except that they utilize larger mesh (10 m long and 4 m wide) (Schemenauer and Cereceda 1994a). A gutter is attached to the lower edge of the mesh which is 2 m above the ground for efficient water collection.

A suitable site for the fog water collection project is selected based on various geographical factors such as elevation, wind pattern, relief in the surrounding area, wind speed, mountain range, availability of a large area for collectors, and slope orientation (Schemenauer and Cereceda 1994a). Once the suitable site is finalized, an exploratory study is conducted for the measurement of fog water collection on daily basis (Kogan and Trahtman 2003). Over the period, several fog collection projects have been successfully carried out in different countries, which is practical proof of their success as a sustainable technology for domestic and agricultural purposes (Table 7.1).

7.2.2 *Dew Water Collection*

7.2.2.1 Dew

Dew is formed by the condensation of water droplets present in the atmosphere on substrates at or near the ground (Monteith 1957). Also, dew is different from distillation, which results from the heat transfer from soil moisture (Monteith 1957). Dew formation comprises the following physical processes: heterogeneous nucleation, drop fusion, renucleation, and drop removal (Beysens et al. 1991; Beysens 2006). This cycle is possible only if the temperature of the surface of the substrate falls below its dew point temperature. This can be achieved at night when there is a lower surface temperature as compared to daytime. Other conditions that favour the formation of dew are the high humidity near the ground and the presence of moisture-laden wind (Tomaszkiewicz et al. 2015; Beysens et al. 1991; Beysens 2006).

Table 7.1 Summary of some of the fog water collection studies in the world

S. No.	Name of country	Site	Reason for study	Average water production (L/m^2/day)	References
1	Iran	Khorassan	To obtain water for forestry, cattle raising, agricultural and domestic use	0.53	Mousavi-baygi (2008)
2	Spain	Canary Islands	To provide water to the local population	10	Marzol et al. (2011)
3	Oman	Dhofar	For drinking purposes and reforestation	20.69	Abdul-Wahab et al. (2007)
4	South Africa	Soutpansber Lepelfontein	For community use	2 4.6	Macedonio et al. (2012)
5	The Kingdom of Saudi Arabia	Asir region	To provide water for agriculture use in remote areas	5.5 (using local mesh); 6.7 (using imported mesh)	Algarni (2018)

7.2.2.2 Dew Collection

Dew water collection is a more suitable water harvesting technique as it is less dependent on climatic and geographical factors as compared to fog water collection. Dew formation generally occurs everywhere and occurs more frequently in comparison to the fog that occurs in very particular locations, and hence dew water collection is more accessible and attractive (Nikolayev et al. 1996). Dew water collection processes are carried out (Tomaszkiewicz et al. 2015; Tu et al. 2018b; Beysens and Milimouk 2000);

(i) Using passive (radiative) cooling condenser,
(ii) Using active cooling condensation technology, and
(iii) Using solar-regenerated desiccant.

Passive cooling condensers utilize physical processes of dew formation to harvest dew water and so do not require external energy input (Tu et al. 2018b). The theoretical basis for the development of efficient radiative condensers was given by Nikolayev and his co-workers (Nikolayev et al. 1996). According to them, an “ideal” condenser is a light sheet that should be placed in an open area to irradiate the energy to space and should be thermally isolated from the ground to avoid the greenhouse effect (Nikolayev et al. 1996; Tu et al. 2018b). Such condensers work well at night when the surface temperature is low. So, clear nights are most advantageous for dew formation in contrast to cloudy nights as it permits greater cooling of surface temperature. These sheets should ideally be of a material that is easily wetted by water to lower the nucleation barrier at the start of the condensation process. Several studies have been carried out to analyze different dew collectors using a passive cooling condenser (Ernesto and Jasson 2016; Khalil et al. 2016; Sharan 2011). The majority of these collectors are made up of flexible material fixed to an inclined surface, from where dew water runs down by gravity to a collector (Khalil et al. 2016; Gerasopoulos et al. 2018). Several studies have been carried out to improve dew water collection by focusing on the condenser material, architecture, and meteorological parameters (Gerasopoulos et al. 2018; Nilsson 1996; Nilsson et al. 1994; Alnaser and Barakat 2000; Silva et al. 2022).

The dependency of passive condensers on specific environmental conditions and low yields leads to an innovative option of active condensers. Active condensers are less dependent on variation in sky emissivity, wind speed, and topographic cover than passive condensers and hence can be utilized under a diverse range of weather conditions (Khalil et al. 2016). Active condensers harvest a significantly larger amount of water than passive condensers but at the expense of additional energy input (Tu et al. 2018b). Despite this drawback, it is used as a supplementary source of water in areas with a high shortage of water. These condensers use cooling condensation technology to harvest water and are suitable both for domestic (15–50 l/day) as well as industrial use (2000 l/day) depending on their design (Taylor et al. 2013).

The major concern in using active condensers is high energy consumption. To reduce the use of fossil fuels, solar regenerated desiccants are preferred. It utilizes desiccants such as zeolites, silica gel, and $CaCl_2$ to capture moisture from the air at night and desorbs water by heating the desiccant using solar radiation. The water vapours thus produced will be liquefied and collected in storage tanks (Taylor et al. 2013; Tu et al. 2018b; Khalil et al. 2016). Liquid desiccants are preferred over solid desiccants owing to their higher moisture-holding capacity and lower regeneration temperature (William et al. 2015). The hygroscopic nature of desiccants enables them to capture and retain a large amount of water and hence such an absorption/adsorption-regeneration system is of special interest, especially in remote areas.

7.3 Understanding Moisture Sorption Mechanism

Water harvesting materials deliver a novel method of addressing the aforementioned difficulties by permitting users to manage the interaction between water molecules and functional materials by different modification knobs. The water affinity provided by the components is often used in materials-abled moisture harvesting. Rather than cooling the surface to enhance local RH, moisture harvesting (MH) materials exploit spontaneous vapor sorption to incarcerate water, collecting vapour from the air and thus concentrating moisture (Schemenauer 2014; Kalmutzki et al. 2018). Through adsorption or/and absorption, they can capture water in both low and saturated RH conditions. The harvested water may be gathered successfully, owing to the necessary accretion of moisture, attributable to the release method being powered by energy inputs for example heating (Fig. 7.1). Owing to the excessive attraction between harvesters and water, one of the most significant issues is the elevated energy utilization for water release. While traditional adsorbents, such as silica gels (Ng et al. 2001; Wang et al. 2005), zeolites (Jänchen et al. 2004; Trzpit et al. 2007), and hygroscopic materials (Kallenberger and Fröba 2018) have proven beneficial, novel concepts based on specific properties of nanostructured materials, such as MOF-based materials (Kalmutzki et al. 2018; Bui et al. 2017b; Kim et al. 2018) and hydrogels (Zhao et al. 2017, 2018, 2019; Li et al. 2018; Entezari et al. 2020; Matsumoto et al. 2018) have also shown promise. High water intake, low energy consumption for water release, excellent cycling stability, speedy water capturing/releasing, high cycling stability, and cost-effectiveness are the fundamental challenges that can be addressed simultaneously by structural designs, and rational materials which have yet to be completely shown. Experiments and theoretical modelling both are necessary for a fundamental understanding of the interaction among materials and moisture content. As a result, the strategies for designing moisture-harvesting materials for AWH are presented in this chapter. The traditional characteristics of humidity capture are described first, as well as relevant underlying mechanisms and essential design ideas. The use of innovative materials and structural designs, as well as illustrative examples, could be utilized to introduce current difficulties and alternatives. AWH based on the moisture harvesting process absorbs vapour from the air via adsorption or, absorption which refers to the physical or chemical addition of water molecules to materials (Cao et al. 2012). AWH relying on the moisture harvesting process gathers vapour from the air via adsorption or, absorption which refers to the physical or chemical adsorption of molecules of water to materials as represented in Fig. 7.2 (LaPotin et al. 2019; Cao et al. 2012). Absorption is a bulk phenomenon in which liquid/gas molecules diffuse through solid/liquid materials, causing the volume and structure of the absorbents to change. Chemical reactions and physical interactions with absorbent materials can absorb water molecules (Sant 2013). Chemical absorption is determined by the reaction's stoichiometry and the concentration of the substrate, whereas physical absorption is mainly influenced by the osmotic pressure. Typical absorption-based moisture harvesters are water-absorbing materials that absorb water molecules through a hydration process that involves both chemical and physical absorption (Kallenberger and Fröba 2018; Entezari et al. 2020; Zhou et al. 2020). When the vapour pressure of confined water is lesser than the partial pressure of vapor in the air with high water absorption, such deliquescent salts can absorb moisture. Though, particle agglomerate during the dehydration/hydration of solid salts may cause performance degradation because of lower water vapour permeability (LaPotin et al. 2019). Furthermore, because of the high vaporization enthalpy, the desorption processes for hygroscopic materials are called a bottleneck (Wang et al. 2019; Qi et al. 2019). Adsorption is a surface process in which liquid or gas molecules attach to a surface via physical (Physisorption) and chemical (chemisorption) interactions. Adsorption thermodynamics are significant to intrinsic material properties. The sorbent surface requires requisite sites to adsorbed molecules via chemical bonding, for example electrostatic interactions, hydrogen bonding, and coordination effect for chemical adsorption.

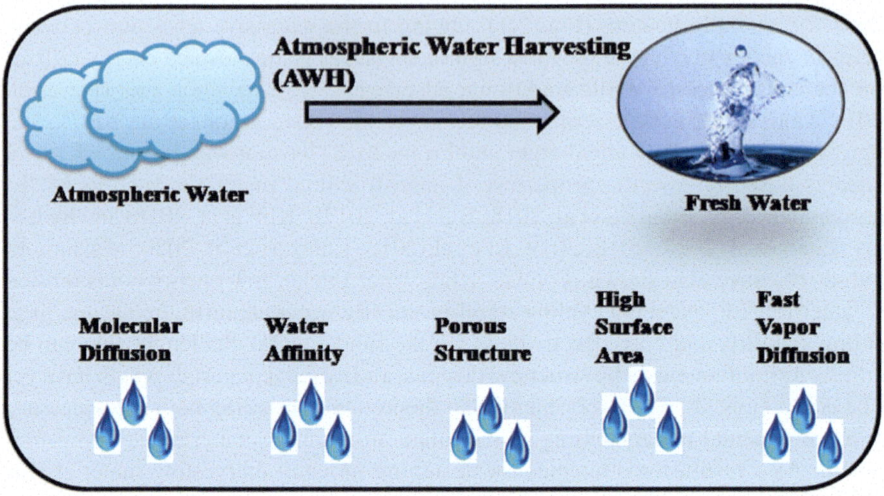

Fig. 7.1 Schematic illustration of AWH depends on moisture harvesters with a wide surface area, water affinity, porous structure, molecular diffusion, and rapid vapor diffusion that can capture water vapour in the atmosphere as fresh water

7.4 Structural Design of AWH

Moisture harvester-enabled AWH technologies must have the following characteristics: moisture capture capability, sorption and desorption kinetics, water release energy demand, and cycling durability. Improved moisture sorption capacity is derived from materials with high porosity, superior water affinity, and huge surface area that can enhance water uptake, harvesting adequate moisture for vapour liquefaction and delivering recoverable water (Gado et al. 2022). An ambient moisture harvester with low generation energy demand, high-water retention, rapid desorption/sorption, and excellent stability can be attained for effective AWH through a suitable selection of materials and reasonable modification (Kalmutzki et al. 2018; LaPotin et al. 2019).

7.4.1 Silica Gel

Silica gel absorbs water uses in a variety of applications. Silica gel as a desiccant is utilized for storage, capture, cooling, and desalination. Silica gel is a desiccant essential in the sorption of water from the atmosphere because it has a low regeneration temperature. Silica gel's sorption performance in AWH technologies is insufficient at ambient temperature, reaching 40%. Essa et al. (2020) employed a half-cylindrical basin solar and dual slope with hygroscopic silica gel to examine water production. A novel highly hygroscopic silica gel was used. Daily water production was

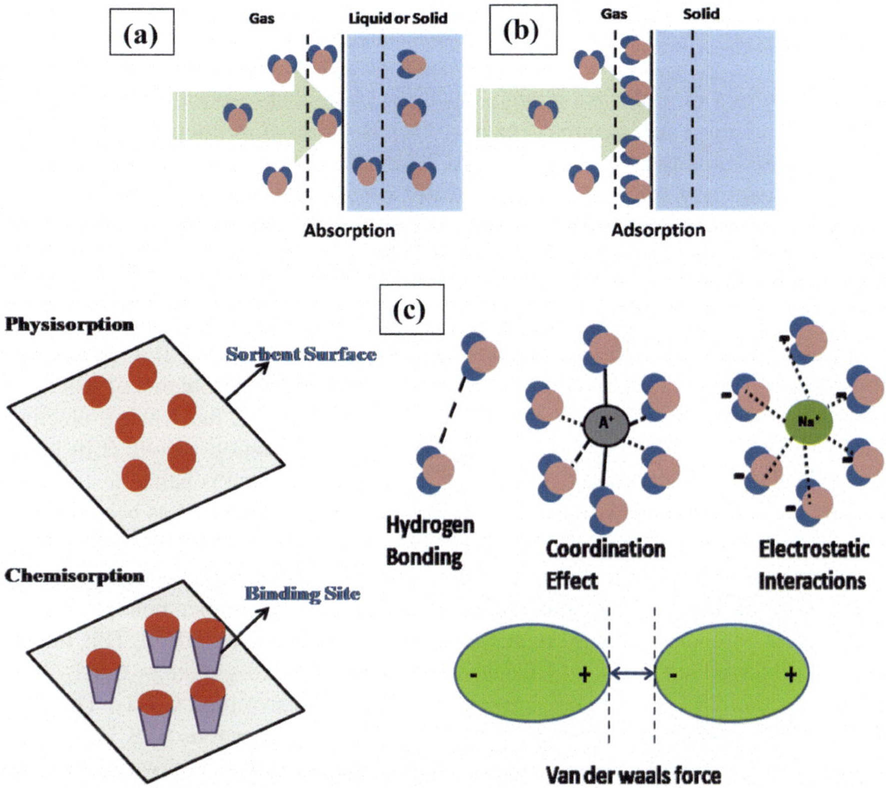

Fig. 7.2 Schematic representation of sorption mechanisms **a** Absorption, **b** adsorption, **c** physisorption/chemisorption process

400 mLm^{-2}, according to their results (Essa et al. 2020). Sleiti et al. (2021) introduced new water collecting concept that used silica gel with a thickness of 25–35 mm as a desiccant material. During a 12-h cycle, it was observed that the suggested system produced 159 g of water per 1 kg of silica gel. Silica gel has limited sorption and thermal capacity resulting in difficulty in the desorption of water (Yuan et al. 2016). Furthermore, the adsorption capability of pure silica gel is less than 40% by weight. As a result, adding additives to increase adsorption capacity is essential (Chen et al. 2010; Daou et al. 2008). Silica gel was improved by adding plentiful additives like hygroscopic salts and metal ions. Doping silica gel with metal ions (e.g., Ti) to enhance its sorption is expensive and has severe environmental implications.

7.4.2 Activated Carbon Fibres

ACF consists of a short diffusion range, resulting in rapid absorption rates, as well as a 1380 m^2/g average surface area (Suzuki 1994). Because of its microstructures with pore diameters ranging from 0.1 to 2 nm, ACF served as a host for several salts leading to a strong capillary force that collects water vapour from atmospheric air. When compared to other salts, the use of $CaCl_2$ as a host material yielded shows potential results. $CaCl_2$ is therefore a better host material for water uptake as compared to silica gel and ACF (Wang et al. 2016). Furthermore, both hosts have a $CaCl_2$composition of 30 wt%. According to their findings, the water absorption capacity is 1.6 g/g of ACF/$CaCl_2$, which is roughly three times i.e., 0.5 g/g at 70% RH and 30 °C adsorption temperature of silica gel/$CaCl_2$. ACF is used to overcome the problems of silica gel carryover and rupture. Under the temp 25 °C and 75% RH, the water collecting unit can yield 1.41 $g_{H_2O}/g_{AS5Li30}$, according to the estimation results (Liu et al. 2021). Furthermore, a corrugated ACF/LiCl composite sorbent is being designed to evaluate its water uptake (Wang et al. 2017). ACF/LiCl + $MgSO_4$ has been used as a composite absorbent and is investigated for water uptake in arid climates with a maximum RH of 35% (Ejeian et al. 2020). The model has a freshwater uptake of 0.92 g/g, according to the results. ACF is combined with another binary solution to produce a composite sorbent ($CaCl_2$ + LiCl/ACF) (Liu et al. 2017). This binary salt was preferred since LiCl/ACF has a higher uptake (2.5 g/g) than $CaCl_2$/ACF (1.7 g/g), although LiCl is nearly ten times more expensive than $CaCl_2$. The host matrix is ACF, and the two salts are utilized in different proportions ($CaCl_{2:}$ LiCl: = 1:3). At temp 25 °C and relative humidity of 90%, the amount of freshwater captured is 2.99 g/g. The composite sorbent has been used to investigate water absorption using ACF as a host material of three separate salts: $LiNO_3$, $CaCl_2$&LiCl. Each salt carries a 20% concentration. ACF/LiCl had the maximum water absorption, measuring 2.9 gg^{-1} at 25 °C and % RH and 1.2 gg^{-1} at 25 °C and % RH. The total quantity of water absorbed by an ACF mixture including Cerium (4% wt) and $MgCl_2$ (40%) at 70% RH and 25 °C is 1.05 gg^{-1}.

7.4.3 Metal Organic Frameworks

MOFs are compounds that build open crystalline frameworks with persistent porosity by associating metal units with organic linkers. MOFs are also known as metal–organic frameworks (Gordeeva et al. 2021). The ideal MOF should possess various characteristics, including (1) considerable water uptake at high relative humidity and low temperatures (2) minimal toxicity, (3) appropriate cycling processes, (4) structural variation, and (5) high adsorption rates (Mouchaham et al. 2020). As a result, these materials can be utilized to be after that generation for water harvesting materials. An innovative investigation was conducted with 23 desiccant materials, 20 of which are MOFs (Furukawa et al. 2014). Furthermore, because the MOF ~ (801-P) and MOF ~ (841) are water stable, they provide the best overall performance for other materials. MOF:74, which can absorb up to 550 cm^3kg^{-1} of water at a relative pressure of 0.1, is potentially an effective desiccant material; however, PIZOF-2 is not ideal for AWH purposes. In the laboratory, MOF ~ (801) is used in an AWH system following the parameters that have been established (Bui et al. 2017b). When the absorption circumstance RH of 20% attempt 35 °C and the water creation is 2.8 Lkg^{-1} for a single daily cycle, it was discovered that the water creation is 2.8 Lkg^{-1} at 20% RH. In arid climates (10–40% RH), the AWH system produces water using MOF ~ (801) is found to be 0.25 Lkg^{-1} or 0.34 Lm^{-2} (Kim et al. 2018). It is important to note that the main disadvantages of employing MOFs in AWH are their expensive cost and poor hydrothermal stability. As a result, additional work is expended in building new MOFs to address those challenges. As, Kim et al. designed and synthesized a novel MOF ~ (801) with superior water uptake and superior hydrothermal stability at substantially lower humidity (i.e., adsorption capability of 0.28 kg/g). Aluminium-based MOFs have the potential to be more cost-effective than their rivals in terms of synthesis, as well as having the ability to perform up to 150 desorption/adsorption cycles daily (Hanikel et al. 2019). Yilmaz et al. (2020) investigated the use of MOFs to improve the performance of a hydro-active polymer (Yilmaz et al. 2020). The blending of a super absorbent with MOFs was also examined by the same scientists to increase adsorption capacity and provide rapid kinetics. Previous findings have revealed further insights on scalable and continuous AWH systems for additional study on the thermal implications of MOFs. MOF ~ (303), MIL ~ 101(Cr), and MOF ~ (841) are the majority of appropriate MOFs, according to previous findings (Gordeeva et al. 2021).

7.4.4 3-D Polymeric Network Hydrogels

A competent adsorbent material is a hydrogel or SMAG (super moisture-absorbent gel), which is made of moisture-absorbing polypyrrole chloride infused in a tuneable hydrophilic network of poly *N*-isopropyl acrylamide as represented in Fig. 7.3 (Zhao et al. 2019). SMAG has demonstrated its ability to capture vapour, liquefy water in situ, store high-density water, and release water quickly in a variety of climatic circumstances. Zhao et al. (2019) fashioned and tested a basic water harvester that can estimate everyday water production. By integrating effective moisture collecting, in situ water liquefaction, huge water storage ability, and rapid water release below a variety of weather conditions, this moisture harvester can achieve high-competence water production in a wide RH range. Furthermore, for composite (Alg-$CaCl_2$), hydrogel hosts the water adsorption (Kallenberger and Fröba 2018). At a temperature of 28 °C and 26% relative humidity (RH), the water uptake exceeds 1 gg^{-1}, according to the findings. Water intake, in this case, can be as high as 660 kg $m^{-3}$of composite material. $CaCl_2$ is used to explore the water generation of cellulose/graphene oxide composite-based aerogels (graphene oxide-based aerogels) (Wang et al. 2019).

During the day, the graphene oxide-based aerogel acts as a salt-anti absorber, while at night, the $CaCl_2$ solution acts as a sorbent material. At a relative humidity of 70%, daily water output can reach 2.89 kgm^{-2}. Additional efforts to promote freshwater production utilizing photo thermal organogel (POG) have been made (Ni et al. 2020). They primarily looked at daily water production in open-air conditions and discovered that absorption and water sorption is around 2.34 and 4.36 kgm^{-2}, respectively. Entezari et al. (2020) used polymeric networks with FCNTs to increase solar spectrum absorption. This composition is strengthened in LiCl and $CaCl_2$ solution. Their research shows that this composite material can absorb 5.6 gg^{-1} of water (Entezari et al. 2020). Deliquescent salts and other mixes that perform well with biomaterials enhance adsorption capability; however, it dissolves in the adsorbed water, restricting their applicability in AWH systems. As a result, for improved adsorption, Mittal et al. (2020) employed acrylic acid and sodium acrylate (AA + SA) highly porous hydrogels incorporated $CaCl_2$ in their composite. $CaCl_2$ does not dissolve and does not influence the polymer matrix. Furthermore, with acceptable adsorption effectiveness, the adsorbent material chosen can work for up to 10 consecutive adsorption/desorption processes (Mittal et al. 2020). The salting-in effect of a zwitterionic hydrogel is shown in this study for the first time to be capable of facilitating water vapour sorption by the hygroscopic salt under otherwise identical conditions. For the demonstration, a salt-hydrogel composite was made by embedding a hygroscopic salt of LiCl into a poly-[2-(methacryloyloxy)ethyl]dimethyl-(3-sulfopropyl)ammonium hydroxide (PDMAPS) zwitterionic hydrogel. The zwitterionic hydrogel matrix has increased swelling capacity as a result of the salting-in effect, which contributes to the high AWH performance of the sorbent (Aleid et al. 2022). Here, Wang et al. established a hydrogel-based composite sorbent that is expandable by confining LiCl into CNTs and acrylamide monomer (AM), short for PCLG, which is equipped with both ultra-high adsorption/desorption capacity per unit volume and low-temperature desorption. They then invented an AWH device with an optimised hydrogel honeycomb structure and demonstrated its water-harvesting performance under natural sunlight at a cooling temperature of 35 °C. Systematic research, from PCLG material and structure optimization to experimental validation, supported our hydrogel AWH design's high water-extraction efficiency (Wang et al. 2022).

7.5 Summary

In this chapter, we discussed the diverse atmospheric water harvesting technologies of fog as well vapours in detail, their working and utilization. Fog water collection is low-cost, low maintenance, and sustainable method in areas which receive scanty rainfall and where fog formation is a regular episode. It utilizes simple technology to extract moisture from the fog by using vertical structures consisting of mesh. Different types of fog collectors result in variable quantities of fog collection depending upon the size, shape, material of mesh utilized, the surface area of the collector, and wind speed. Fog collectors are classified as Standard fog collectors

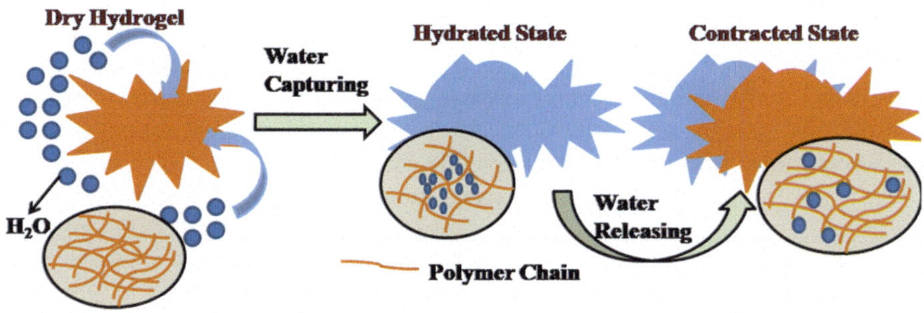

Fig. 7.3 Atmospheric rainwater harvesting (AWH) based on the super moisture absorbent gel (SMAG)

(SFCs) and large fog collectors (LFCs). Another technique, Dew water collection is a more suitable water harvesting technique as it is less dependent on climatic and geographical factors as compared to fog water collection. Passive cooling condensers utilize physical processes of dew formation to harvest dew water and so do not require external energy input. Water harvesting resources provide an innovative tactic for addressing the above-mentioned complications by allowing users to control the interaction between functional materials and water molecules using different tuning knobs. The strategies for designing moisture-harvesting materials for AWH are presented in this chapter. The use of innovative materials and structural designs could be utilized to introduce current difficulties and alternatives. AWH based on the moisture harvesting process absorbs vapour from the air via absorption or adsorption, which refers to the physical or chemical attachment of water molecules to materials. Moisture harvester-enabled AWH technologies must have the following characteristics: moisture capture capacity, water release energy demand, and sorption and desorption kinetics, and cycling durability.

In this respect, silica gel is widely used for storing, capturing, cooling, and desalination. When utilized singly in AWH technologies, however, silica gel's sorption performance is insufficient at ambient temperature and reaches 40%. Activated carbon fibre (ACF) is used to overcome the problems of silica gel carryover and rupture. Metals are joined together with organic components to form open frameworks with enduring absorbency, known as MOFs (Metal–Organic Frameworks). These materials can be utilized to be the next generation as water harvesting materials. A capable adsorbent substantial is hydrogel or super moisture-absorbent gel (SMAG). It has demonstrated its ability to capture vapour, liquefy water in situ, store high-density water, and release water quickly in a variety of climatic circumstances.

Future Scope

In recent years, plentiful studies engrossed in the advancement of hydrogels have been stated in the literature. The achievement of these soft constituents can be attributed to their several attractive possessions. Such polymers have a system that, due to their hydrophilic nature, may absorb huge quantities of water and physiological fluids without dissolving. The constituents are most popular due to their thermo- and pH sensitivity, which allows the proposal of smart materials. That's why these constituents can be enthused in their usage in a wide range of applications (biomaterials, delivery systems, soil conditioners, environmental remediation, among others). Thermo sensitive polymers can be compatible with extrusion-based 3D printing also which requires substantial viscosity change once the polymer is extruded onto a substrate.

References

Abdul-Wahab SA, Lea V (2008) Reviewing fog water collection worldwide and in Oman. Int J Environ Stud 65:487–500. https://doi.org/10.1080/00207230802149983

Abdul-Wahab SA, Al-Hinai H, Al-Najar KA, Al-Kalbani MS (2007) Feasibility of fog water collection: a case study from Oman. J Water Supply Res Technol 56:275–280. https://doi.org/10.2166/aqua.2007.045

Aleid S, Wu M, Li R, Wang W, Zhang C, Zhang L, Wang P (2022) Salting-in effect of zwitterionic polymer hydrogel facilitates atmospheric water harvesting. ACS Mater Lett 4:511–20. https://doi.org/10.1021/acsmaterialslett.1c00723

Algarni S (2018) Assessment of fog collection as a sustainable water resource in the southwest of the Kingdom of Saudi Arabia. Water Environ J 32:301–309. https://doi.org/10.1111/wej.12330

Alnaser WE, Barakat A (2000) Use of condensed water vapour from the atmosphere for irrigation in Bahrain. Appl Energy 65:3–18. https://doi.org/10.1016/S0306-2619(99)00054-9

Alqadami AA, Naushad M, Alothman ZA, Ghfar AA (2017) Novel metal-organic framework (MOF) based composite material for the sequestration of U(VI) and Th(IV) metal ions from aqueous environment. ACS Appl Mater Interfaces 9:36026–36037. https://doi.org/10.1021/acsami.7b10768

Batisha AF (2015) Sustainability of water quality and ecology feasibility and sustainability of fog harvesting 6:5–7

Beysens D (2006) Dew nucleation and growth. Comptes Rendus Phys 7:1082–1100

Beysens D, Milimouk I (2000) The case for alternative fresh water sources. Pour Les Resour Altern En Eau, Secher 11:1–17

Beysens D, Steyer A, Guenoun P, Fritter D, Knobler CM (1991) How does dew form? Phase Transitions 31:219–246. https://doi.org/10.1080/01411599108206932

Bui DT, Chua KJ, Gordon JM (2017a) Comment on "Water harvesting from air with metal-organic frameworks powered by natural sunlight" 0791:1–3

Bui DT, Chua KJ, Gordon JM (2017b) Comment on "Water harvesting from air with metal-organic frameworks powered by natural sunlight.". Science (80-) 358:1–10. https://doi.org/10.1126/science.aao0791

Cao Y, Chen Y, Sun X, Zhang Z, Mu T (2012) Water sorption in ionic liquids: Kinetics, mechanisms and hydrophilicity. Phys Chem Chem Phys 14:12252–12262. https://doi.org/10.1039/c2cp41798g

Chen CJ, Wang RZ, Xia ZZ, Kiplagat JK, Lu ZS (2010) Study on a compact silica gel-water adsorption chiller without vacuum valves: design and experimental study. Appl Energy 87:2673–2681. https://doi.org/10.1016/j.apenergy.2010.03.022

Da Silva LC, Filho DO, Imbuzeiro HMA, Monteiro PMB (2022) Analysis of different condensing surfaces for dew harvesting. Water Supply 22:697–706. https://doi.org/10.2166/ws.2021.242

Daou K, Wang RZ, Yang GZ, Xia ZZ (2008) Theoretical comparison of the refrigerating performances of a $CaCl_2$ impregnated composite adsorbent to those of the host silica gel. Int J Therm Sci 47:68–75. https://doi.org/10.1016/j.ijthermalsci.2007.01.003

Domen JK, Stringfellow WT, Camarillo MK, Gulati S (2014) Fog water as an alternative and sustainable water resource. Clean Technol Environ Policy 16:235–249. https://doi.org/10.1007/s10098-013-0645-z

Ejeian M, Entezari A, Wang RZ (2020) Solar powered atmospheric water harvesting with enhanced LiCl /MgSO4/ACF composite. Appl Therm Eng 176:115396. https://doi.org/10.1016/j.applthermaleng.2020.115396

Entezari A, Ejeian M, Wang R (2020) Super atmospheric water harvesting hydrogel with alginate chains modified with binary salts. ACS Mater Lett 2:471–477. https://doi.org/10.1021/acsmaterialslett.9b00315

Ernesto ATJ, Jasson FPJ (2016) Winter dew harvest in Mexico City. Atmosphere (basel) 7:1–12. https://doi.org/10.3390/atmos7010002

Essa FA, Elsheikh AH, Sathyamurthy R, Muthu Manokar A, Kandeal AW, Shanmugan S, Kabeel AE, Sharshir SW, Panchal H, Younes MM (2020) Extracting water content from the ambient air in a double-slope half-cylindrical basin solar still using silica gel under Egyptian conditions. Sustain Energy Technol Assess 39:100712. https://doi.org/10.1016/j.seta.2020.100712

Fath H, Abbas Z, Khaled A (2011) Techno-economic assessment and environmental impacts of desalination technologies 266:263–273. https://doi.org/10.1016/j.desal.2010.08.035

Fessehaye M, Abdul-Wahab SA, Savage MJ, Kohler T, Gherezghiher T, Hurni H (2014) Fog-water collection for community use. Renew Sustain Energy Rev 29:52–62. https://doi.org/10.1016/j.rser.2013.08.063

Furukawa H, Gándara F, Zhang YB, Jiang J, Queen WL, Hudson MR, Yaghi OM (2014) Water adsorption in porous metal-organic frameworks and related materials. J Am Chem Soc 136:4369–4381. https://doi.org/10.1021/ja500330a

Gado MG, Nasser M, Hassan AA, Hassan H (2022) Adsorption-based atmospheric water harvesting powered by solar energy: comprehensive review on desiccant materials and systems. Process Saf Environ Prot 160:166–183. https://doi.org/10.1016/j.psep.2022.01.061

Gerasopoulos K, Luedeman WL, Ölçeroglu E, McCarthy M, Benkoski JJ (2018) Effects of engineered wettability on the efficiency of dew collection. ACS Appl Mater Interfaces 10:4066–4076. https://doi.org/10.1021/acsami.7b16379

Gido B, Friedler E, Broday DM (2016) Assessment of atmospheric moisture harvesting by direct cooling. Atmos Res 182:156–162. https://doi.org/10.1016/j.atmosres.2016.07.029

Gordeeva LG, Tu YD, Pan Q, Palash ML, Saha BB, Aristov YI, Wang RZ (2021) Metal-organic frameworks for energy conversion and water harvesting: a bridge between thermal engineering and material science. Nano Energy 84:105946. https://doi.org/10.1016/j.nanoen.2021.105946

Hanikel N, Prévot MS, Fathieh F, Kapustin EA, Lyu H, Wang H, Diercks NJ, Glover TG, Yaghi OM (2019) Rapid cycling and exceptional yield in a metal-organic framework water harvester. ACS Cent Sci 5:1699–1706. https://doi.org/10.1021/acscentsci.9b00745

Jänchen J, Ackermann D, Stach H, Brösicke W (2004) Studies of the water adsorption on Zeolites and modified mesoporous materials for seasonal storage of solar heat. Sol Energy 76:339–344. https://doi.org/10.1016/j.solener.2003.07.036

Jarimi H, Powell R, Riffat S (2020) Review of sustainable methods for atmospheric water harvesting. Int J Low-Carbon Technol 15:253–276. https://doi.org/10.1093/ijlct/ctz072

Kallenberger PA, Fröba M (2018) Water harvesting from air with a hygroscopic salt in a hydrogel–derived matrix. Commun Chem 1:6–11. https://doi.org/10.1038/s42004-018-0028-9

Kalmutzki MJ, Diercks CS, Yaghi OM (2018) Metal-organic frameworks for water harvesting from air. Adv Mater 30:1–26, 1704304. https://doi.org/10.1002/adma.201704304

Khalil B, Adamowski J, Shabbir A, Jang C, Rojas M, Reilly K, Ozga-Zielinski B (2016) A review: dew water collection from radiative passive collectors to recent developments of active collectors. Sustain Water Resour Manag 2:71–86. https://doi.org/10.1007/s40899-015-0038-z

Kim H, Rao SR, Kapustin EA, Zhao L, Yang S, Yaghi OM, Wang EN (2018) Adsorption-based atmospheric water harvesting device for arid climates. Nat Commun 9:1–8. https://doi.org/10.1038/s41467-018-03162-7

Klemm O, Schemenauer RS, Lummerich A, Cereceda P, Marzol V, Corell D, Van Heerden J, Reinhard D, Gherezghiher T, Olivier J, Osses P, Sarsour J, Frost E, Estrela MJ, Valiente JA, Fessehaye GM (2012) Fog as a fresh-water resource: overview and perspectives. Ambio 41:221–234. https://doi.org/10.1007/s13280-012-0247-8

Klemm O, Schemenauer RS, Lummerich A, Cereceda P, Marzol V, Corell D, Heerden J Van, Reinhard D, Gherezghiher T, Olivier J (2012b) Fog as a fresh—water resource: overview and perspectives, 1–26

Kogan B, Trahtman A (2003) The moisture from the air as water resource in arid region: hopes , doubts and facts 40:231–240. https://doi.org/10.1006/jare.2002.1028

LaPotin A, Kim H, Rao SR, Wang EN (2019) Adsorption-based atmospheric water harvesting: impact of material and component properties on system-level performance. Acc Chem Res 52:1588–1597. https://doi.org/10.1021/acs.accounts.9b00062

Li R, Shi Y, Alsaedi M, Wu M, Shi L, Wang P (2018) Hybrid hydrogel with high water vapor harvesting capacity for deployable solar-driven atmospheric water generator. Environ Sci Technol 52:11367–11377. https://doi.org/10.1021/acs.est.8b02852

Liu JY, Wang JY, Wang LW, Wang RZ (2017) Étude Expérimentale Sur Les Propriétés Des Sorbants Composites Pour Des Paires Actives À Sorption D'Eau Triphasique. Int J Refrig 83:51–59. https://doi.org/10.1016/j.ijrefrig.2017.07.007

Liu XY, Wang WW, Xie ST, Pan QW (2021) Performance characterization and application of composite adsorbent LiCl@ACFF for moisture harvesting. Sci Rep 11.https://doi.org/10.1038/s41598-021-93784-7

Macedonio F, Drioli E, Gusev AA, Bardow A, Semiat R, Kurihara M (2012) Chemical engineering and processing: process intensification efficient technologies for worldwide clean water supply. Chem Eng Process Process Intensif 51:2–17. https://doi.org/10.1016/j.cep.2011.09.011

Malik FT, Clement RM, Gethin DT, Krawszik W, Parker AR (2014) Nature's moisture harvesters: a comparative review. 031002. https://doi.org/10.1088/1748-3182/9/3/031002

Marzol V, Sánchez JL, Yanes A (2011) Meteorological patterns and fog water collection in morocco and the canary Islands. Erdkunde 65:291–303. https://doi.org/10.3112/erdkunde.2011.03.06

Matsumoto K, Sakikawa N, Miyata T (2018) Thermo-responsive gels that absorb moisture and ooze water. Nat Commun 9:1–7. https://doi.org/10.1038/s41467-018-04810-8

Mittal H, Al Alili A, Alhassan SM (2020) Adsorption isotherm and kinetics of water vapors on novel superporous hydrogel composites. Micropor Mesopor Mater 299:110106. https://doi.org/10.1016/j.micromeso.2020.110106

Monteith JL (1957) Dew Q J R Meteorol Soc 83:322–341. https://doi.org/10.1002/qj.49708335706

Mouchaham G, Cui FS, Nouar F, Pimenta V, Chang JS, Serre C (2020) Metal-organic frameworks and water: 'from old enemies to friends'? Trends Chem 2:990–1003. https://doi.org/10.1016/j.trechm.2020.09.004

Mousavi-baygi M (2008) The implementation of fog water collection systems in Northeast of Iran. Int J Pure Appl Phys 4:13–21

Munoz-garcia MA, Moreda GP (2005) Water harvesting for young trees using Peltier modules powered by photovoltaic solar energy

Ng KC, Chua HT, Chung CY, Loke CH, Kashiwagi T, Akisawa A, Saha BB (2001) Experimental investigation of the silica gel-water adsorption isotherm characteristics. Appl Therm Eng 21:1631–1642. https://doi.org/10.1016/S1359-4311(01)00039-4

Ni F, Qiu N, Xiao P, Zhang C, Jian Y, Liang Y, Xie W, Yan L, Chen T (2020) Tillandsia-inspired hygroscopic photothermal organogels for efficient atmospheric water harvesting. Angew Chemie 132:19399–19408. https://doi.org/10.1002/ange.202007885

Nikolayev VS, Beysens D, Gioda A, Milimouka I, Katiushin E, Morel J-P (1996) Water recovery from dew. J Hydrol 182:19–35. https://doi.org/10.1016/0022-1694(95)02939-7

Nilsson T (1996) Initial experiments on dew collection in Sweden and Tanzania. Sol Energy Mater Sol Cells 40:23–32. https://doi.org/10.1016/0927-0248(95)00076-3

Nilsson TMJ, Vargas WE, Niklasson GA, Granqvist CG (1994) Condensation of water by radiative cooling. Renew Energy 5:310–317. https://doi.org/10.1016/0960-1481(94)90388-3

Olivier J, De RCJ (2002) The implementation of fog water collection systems in South Africa 64:227–238
Park K, Chhatre SS, Srinivasan S, Cohen RE, Mckinley GH (2013) Optimal design of permeable fiber network structures for fog harvesting
Qi H, Wei T, Zhao W, Zhu B, Liu G, Wang P, Lin Z, Wang X, Li X, Zhang X, Zhu J (2019) An interfacial solar-driven atmospheric water generator based on a liquid sorbent with simultaneous adsorption-desorption. Adv Mater 31:1–9. https://doi.org/10.1002/adma.201903378
Sant SB (2013) Phys Chem Interfaces
Schemenauer S (2014) Fog-water collection in arid Coasta Locations 20:303–308
Schemenauer R, Cereceda P (1992) Water from fog-covered mountains. Waterlines 10:10–13. https://doi.org/10.3362/0262-8104.1992.013
Schemenauer RS, Cereceda P (1994a) Fog collection's role in water planning for developing countries. Nat Resour Forum 18:91–100. https://doi.org/10.1111/j.1477-8947.1994.tb00879.x
Schemenauer RS, Cereceda P (1994b) A proposed standard fog collector for use in high-elevation regions. J Appl Meteorol 33:1313–1322. https://doi.org/10.1175/1520-0450(1994)033%3c1313:APSFCF%3e2.0.CO;2
Sharan G (2011) Harvesting dew with radiation cooled condensers to supplement drinking water supply in semi-arid. Int J Serv Learn Eng Humanit Eng Soc Entrep 6:130–150. https://doi.org/10.24908/ijsle.v6i1.3188
Sleiti AK, Al-Khawaja H, Al-Khawaja H., Al-Ali M (2021) Harvesting water from air using adsorption material—Prototype and experimental results_Elsevier Enhanced Reader.pdf
Suzuki M (1994) Activated carbon fiber: fundamentals and applications. Carbon N Y 32:577–586. https://doi.org/10.1016/0008-6223(94)90075-2
Taylor P, Guan H, Sebben M, Bennett J (2013) Radiative- and artificial-cooling enhanced dew collection in a coastal area of South Australia. 37–41. https://doi.org/10.1080/1573062X.2013.765494
Tomaszkiewicz M, Abou Najm M, Beysens D, Alameddine I, El-Fadel M (2015) Dew as a sustainable non-conventional water resource: a critical review. Environ Rev 23:425–442. https://doi.org/10.1139/er-2015-0035
Trzpit M, Soulard M, Patarin J, Desbiens N, Cailliez F, Boutin A, Demachy I, Fuchs AH (2007) The effect of local defects on water adsorption in silicalite-1 zeolite: a joint experimental and molecular simulation study. Langmuir 23:10131–10139. https://doi.org/10.1021/la7011205
Tu Y, Wang R, Zhang Y, Wang J (2018a) Progress and expectation of atmospheric water harvesting_Elsevier Enhanced Reader.pdf
Tu Y, Wang R, Zhang Y, Wang J (2018b) Progress and expectation of atmospheric water harvesting. Joule 2:1452–1475. https://doi.org/10.1016/j.joule.2018.07.015
Vuollekoski H, Vogt M, Sinclair VA, Duplissy J, Järvinen H, Kyrö E, Makkonen R, Petäjä T (2015) Estimates of global dew collection potential on artificial surfaces. 601–613. https://doi.org/10.5194/hess-19-601-2015
Wang DC, Xia ZZ, Wu JY, Wang RZ, Zhai H, Dou WD (2005) Study of a novel silica gel-water adsorption chiller. Part I. Design and performance prediction. Int J Refrig 28:1073–1083. https://doi.org/10.1016/j.ijrefrig.2005.03.001
Wang JY, Wang RZ, Wang LW (2016) Water vapor sorption performance of ACF-CaCl2 and silica gel-CaCl2 composite adsorbents. Appl Therm Eng 100:893–901. https://doi.org/10.1016/j.applthermaleng.2016.02.100
Wang JY, Wang RZ, Wang LW, Liu JY (2017) A high efficient semi-open system for fresh water production from atmosphere. Energy 138:542–551. https://doi.org/10.1016/j.energy.2017.07.106
Wang X, Li X, Liu G, Li J, Hu X, Xu N, Zhao W, Zhu B, Zhu J (2019) An interfacial solar heating assisted liquid sorbent atmospheric water generator. Angew Chemie 131:12182–12186. https://doi.org/10.1002/ange.201905229
Wang J, Deng C, Zhong G, Ying W, Li C, Wang S, Liu Y, Wang R, Zhang H (2022) High-yield and scalable water harvesting of honeycomb hygroscopic polymer driven by natural sunlight. Cell Reports Phys Sci 3:100954.https://doi.org/10.1016/j.xcrp.2022.100954

William GE, Mohamed MH, Fatouh M (2015) Desiccant system for water production from humid air using solar energy. Energy 90:1707–1720. https://doi.org/10.1016/j.energy.2015.06.125

Yilmaz G, Meng FL, Lu W, Abed J, Peh CKN, Gao M, Sargent EH, Ho GW (2020) Autonomous atmospheric water seeping MOF matrix. Sci Adv 6:1–9. https://doi.org/10.1126/sciadv.abc8605

Yuan Y, Zhang H, Yang F, Zhang N, Cao X (2016) Inorganic composite sorbents for water vapor sorption: a research progress. Renew Sustain Energy Rev 54:761–776. https://doi.org/10.1016/j.rser.2015.10.069

Zhao F, Shi Y, Pan L, Yu G (2017) Multifunctional nanostructured conductive polymer gels: synthesis, properties, and applications. Acc Chem Res 50:1734–1743. https://doi.org/10.1021/acs.accounts.7b00191

Zhao F, Bae J, Zhou X, Guo Y, Yu G (2018) Nanostructured functional hydrogels as an emerging platform for advanced energy technologies. Adv Mater 30:1–16. https://doi.org/10.1002/adma.201801796

Zhao F, Zhou X, Liu Y, Shi Y, Dai Y, Yu G (2019) Super moisture-absorbent gels for all-weather atmospheric water harvesting. Adv Mater 31:1–7. https://doi.org/10.1002/adma.201806446

Zhou X, Lu H, Zhao F, Yu G (2020) Atmospheric water harvesting: a review of material and structural designs. ACS Mater Lett 2:671–684. https://doi.org/10.1021/acsmaterialslett.0c00130

Chapter 8
Techno-economic Assessment of Atmospheric Water Harvesting (AWH) Technologies

Ahmed A. Hassan, Mohammed Ezzeddine, Mohamed G. M. Kordy, and Mohamed M. Awad

8.1 Introduction

Sustainable access to freshwater has been recognized as one of the great engineering challenges of the twenty-first century. Freshwater plays vital role in Agriculture, power stations, oil and gas production and all cooling systems. As a result, different technologies for obtaining freshwater from unconventional sources, such as seawater and brackish water desalination, have been developed (Greenlee et al. 2009; Hassan et al. 2022a). Additionally, wastewater reuse and water harvested from the atmosphere, such as rain, fog, and dew harvesting can also be used (Shanyengana et al. 2003; Pimentel-Rodrigues and Silva-Afonso 2022; Guo et al. 2022; Zhang et al. 2022a; Lu et al. 2022). Meanwhile, the economic benefits of fog, rain, and dew harvesting are being addressed and studied. The earth's atmosphere carries water in the form of water droplets or vapor, accounting for up to 10% of freshwater sources and can provide freshwater in the amount of nearly 50,000 km^3 (Zhou et al. 2020). As a result, atmospheric water harvesting (AWH) emerges as a feasible technique for decentralized water production and solving the challenges of long-distance water supply in rural areas. AWH technologies can be classified as illustrated in Fig. 8.1. An excellent AWH system should involve the following characteristics: good water

A. A. Hassan (✉)
Mechanical Power Engineering Department, Zagazig University, Zagazig 44519, Egypt
e-mail: ahmed.adel@zu.edu.eg

M. Ezzeddine · M. M. Awad (✉)
Mechanical Power Engineering Department, Mansoura University, Mansoura 35516, Egypt
e-mail: m_m_awad@mans.edu.eg

M. G. M. Kordy
Nanophotonics and Applications (NPA) Lab, Physics Department, Faculty of Science, Beni-Suef University, Beni-Suef 62514, Egypt

Biochemistry Department, Faculty of Science, Beni-Suef University, Beni-Suef 62521, Egypt

E. Fosso-Kankeu et al. (eds.), *Atmospheric Water Harvesting Development and Challenges*, Water Science and Technology Library 122,
https://doi.org/10.1007/978-3-031-21746-3_8

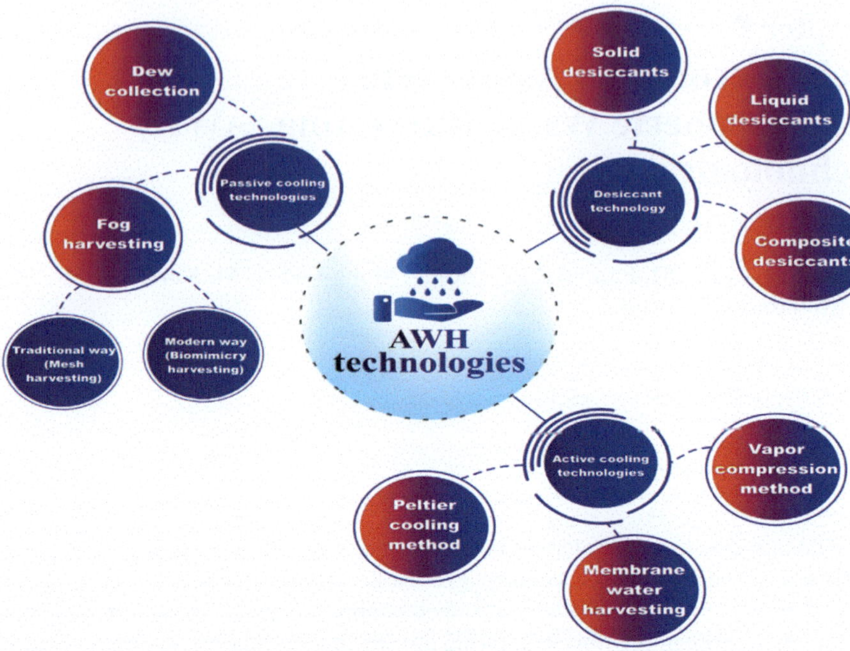

Fig. 8.1 Classification of the main AWH technologies (Bilal et al. 2022)

harvesting capability, including efficient vapor condensation and liquid water collection, low energy consumption, less environmental and climatic limitations, good stability, scalable and robustness (Liu et al. 2022).

Fog water harvesting (FWH) technology's main purpose is to speed up the growth of water droplets attached to the collector surfaces and reduce the critical size of water droplets falling from the surface using surface wettability engineering, or biomimetic micro/nano-structuring (Zhou et al. 2020; Yin et al. 2017; Söz et al. 2020; Parker and Lawrence 2001a; Zheng et al. 2010). FWH technology has the advantages of no power consumption, minimal maintenance cost, and is considered a great freshwater option for arid coastal regions. Artificial rain can be obtained by introducing aerosols into the atmosphere, but cloud seeding is a very expensive technology, and its efficiency is still under evaluation. Dew water harvesting technologies are depending on passing air on cooled surfaces where condensation takes place and water is collected. Surface cooling can be achieved by using vapor compression (VC) cycle or Peltier effect in thermoelectric generators. These technologies consume high electrical power within 0.18–5.21 kWh/kg collected water (Tu and Hwang 2020). Adsorption-based system is promising as it can be operated by renewable sources or waste heat and is not limited by relative humidity of the ambient air. Consequently, adsorption-based AWH systems have become the subject of research by reputable scientific organizations in the past few decades (Gado et al. 2022; Ejeian and Wang 2021). In this chapter, the working principles and the state of the art of each AWH technology are presented.

8.2 AWH Technologies

8.2.1 Artificial Rain Harvesting Technologies

Artificial rain can be a good source of water in areas with high-humidity air. it can be realized in three steps; (1) firstly artificial aerosols of calcium chloride are injected in the air, motivating the condensation process to form cloud droplets, (2) Secondly, aerosols like dry ice and salt are added to accelerate the past process, (3) Finally, silver iodide and dry ice are supplied to improve rainfall. Shukla et al. (2010) proposed a non-linear model for rain from vapor water in the air and his analysis showed clearly that the rainfall increases with the density of the concentrations of the conducive aerosol particles and water vapor. After several investigations on the impact of rain on reducing pollution in different places around the world like England (Davies 1967) and India (Sharma et al. 1967), Shukla et al. (2008) analyzed a nonlinear model for getting rid of pollution in the air and different particulates by rain. Artificial rain can be used in the field of decreasing pollution as a pollution scavenger.

8.2.2 Fog Water Harvesting (FWH) Technologies

There is a shortage in freshwater resources in arid coastal regions which makes the pursuit of capturing vapor water from air with low energy consumption systems without complicated filtration processes an urgent task. Fog Water Harvesting (FWH) systems can play a vital role in these conditions (Zhu et al. 2016). Further the benefits of the FWH systems for freshwater harvesting has a great potential in industrial applications such as waste steam and cooling tower fog harvesting systems (Ghosh et al. 2015). In this section, a summary of fog harvesting technologies inspired by nature and the factors that affect collecting water rates are introduced.

8.2.2.1 Innovation Through Nature

Namib Desert Beetle

Parker and Lawrence (2001b) studied the abilities of the Namib beetle water harvesting method. The water vapor is captured from foggy wind and collected on the hydrophilic bumps then move in hydrophobic grooves toward the mouth of the beetle. Hydrophobic-Hydrophilic contrast triggered special interest for improving the material of water harvesting systems (Wang et al. 2014, 2016a; Gao et al. 2018; Xu et al. 2018).

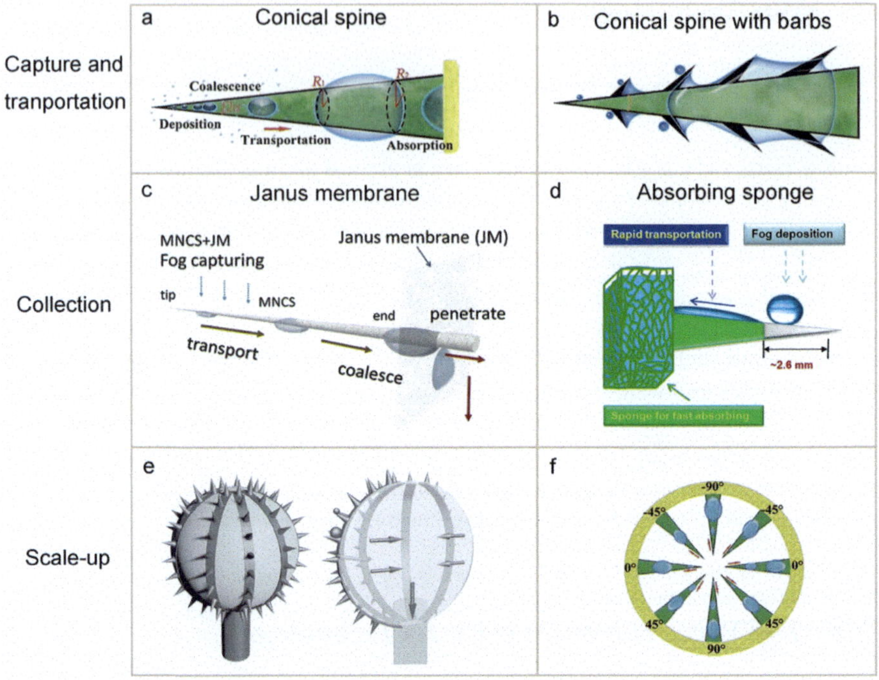

Fig. 8.2 **a** schematic of water collection on a conical spine (Yi et al. 2019), **b** schematic of backward-barbs on conical spine (Yi et al. 2019), **c** schematic of water collection using Janus membrane and a conical spine (Zhou et al. 2018), **d** schematic of absorbing sponge as a storage and a conical spine as harvester (Chen et al. 2018a), **e** schematic of conical spines on the outer area of a spherical cactus-inspired device (Cao et al. 2014), **f** schematic of an inner surface with conical spines inside a spherical water-absorbing surface (Yi et al. 2019)

Cactaceae Species

Cactus in arid areas is known for adapting to drought climates by capturing fog and the transportation of water using its conical spine and gradient in wettability. Many investigations and designs (Cao et al. 2014; Yi et al. 2019; Zhou et al. 2018; Chen et al. 2018a) are proposed inspired by the cactus as shown in Fig. 8.2 (Wang et al. 2021a).

Spider Web

On foggy mornings, everyone can observe the droplets of water collected on spider webs. Following the design of spider silk, Zhai et al. (2010) used silk from nylon and got similar results to the natural spider silk as harvesting capability was the same. more investigations and enhancements were established in Li et al. (2019), Azeem et al. (2020).

There are several biological materials with water harvesting abilities like Sarracenia trichome (Chen et al. 2018b) and syntrichia caninervis captures fog through awns on its leaves (Pan et al. 2016).

8.2.2.2 Significant Factors in the FWH System

To design a good FHW system, there are several factors that have a great impact on the system. Some are climate-dependent such as humidity in the air and wind (Cengel and Boles n.d.; Peng et al. 2015) and the height of the fog harvesting system above sea level (Olivier 2004). Other factors including mesh design, surface roughness, the material of collector, shade coefficient and fog collector orientation as summarized in Fig. 8.3.

There are different patterns used widely like rectangular, woven textile and Rachel in mesh design with every design has its own characteristics and advantages (Fernandez et al. 2018). When the surface is rough meaning that it has many holes and a desire to absorb water, thus less collected water from fog. The shade coefficient is an indication of the ratio between the mesh opening areas to the total area. If the opening area is small, then the shade coefficient is high. It is found that the optimum shade coefficient is in the range of 0.5–0.6 for maximum aerodynamics collection efficiency (Dios Rivera 2011). Orientation of the most existing fog harvesters are standing vertically orthogonal to the direction of the wind. De Dios Rivera (2011) found that concaving mesh induces higher harvesting rates. The surface material is the most critical factor with a great impact on the quantities of collected water. If the material of the surface is attractive to water, it is called hydrophilic material and if it is repulsive to water, it is called hydrophobic material. Also, contact angle plays a crucial role in defining material properties, the higher contact angle means the material is hydrophobic (Good 1992).

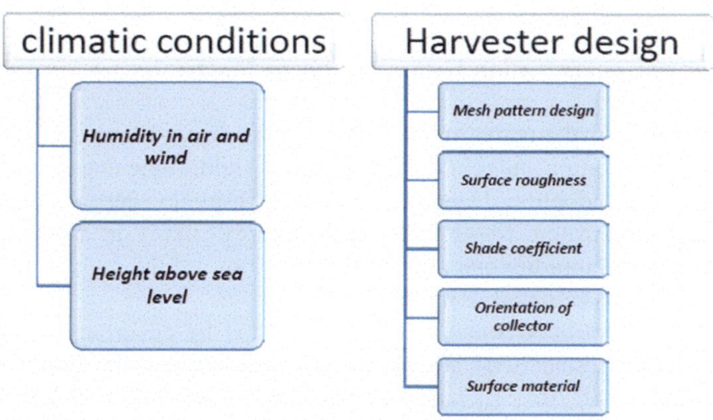

Fig. 8.3 Factors with significant impacts on the FWH system performance

Research on several special materials with favorable characteristics for FHW systems was intensively performed in the recent years. Zhang et al. (2018) succeeded in accelerating water droplet collection rate after coating hydrophilic microfibers with hydrophobic material achieving a double rate of water collection compared with uncoated surfaces and more than five times if the surface is tilted with 5°. Wang et al. (2018a) fabricated hydrophobic elastomer skin from polydimethylsiloxane getting high contact angle of 150°. Using the same idea in the field of FWH, Su et al. (2019) created a smart Janus membrane from a uniaxially stretched polydimethylsiloxane sheet, when strain is 100% the collected water after one hour was 3 ml and when the strain reached 200% the collected water after one hour was 3.9 ml.

Raut et al. (2019) developed a polypropylene hierarchical textured surface that consisted of micro-lenses arrays covered with a concentrated array of clustered fibrils as fog harvester material, also a comparison with the plan surface was recorded. Liu et al. (2019) proposed a hydrophilic-superhydrophobic surface from soy protein achieving water collection rate of 9.17 kg/m^2 h. Pei et al. (2020) proposed a novel coating technique to get an excellent hydrophobic surface using silane-terminated fluorinated polymer, reporting that the contact angle can reach 117°. Sharma et al. (2019) designed CuO nanoneedle surfaces and studied the effect of coating on the performance of collecting water, the coated CuO surface achieved around 750 mL/m^2 h of water with a 165° contact angle.

Generally, the FWH system depends on passive radiation and improving the properties of the harvester to optimize morphology and wettability of the surface and accelerate water nucleation, transportation and collection. Fog harvesters are recommended in high humidity areas, the next demonstrated AWH systems are good choice in less humid areas, but they need power to cool the condenser surface or get captured moisture in Desiccant material free (Liu et al. 2022).

8.2.3 Dew Water Harvesting Technologies

8.2.3.1 Vapor Concentration Using Adsorbent Materials

Despite the merits of the active direct condensation water harvesting techniques, e.g., continuous operation and relatively high-water yield, some main disadvantages still hinder its wide adoption. Depending on the high-grade energy source such as electricity or even burning fossil fuels for its energy supply in addition to their unfeasibility in low humidity and temperatures regions are the main disadvantages of the active methods (Tu et al. 2018).

The vapor concentration water harvesting method by adsorbent materials has attracted numerous researchers' attention in the last few decades due to its many advantages (Gado et al. 2022; Ejeian and Wang 2021). These sorption-based systems can also be used for other applications such as desalination, cooling, and heat pump (Summa et al. 2021; Alsaman et al. 2022; Moossa et al. 2022; Hassan et al. 2020a, b, 2021, 2022b). The main working principle of these water harvesting systems is the sorption-regeneration-condensation method, in which the solid or liquid desiccant material adsorbs water vapor from the humid air, usually at night or during the day, by cooling the material. Then the regeneration process occurs by increasing the temperature of the desiccant material by supplying thermal energy to the material, and finally condensing the water vapor in a condenser and collecting the water by gravity in the collection tank. This method has three main advantages: (i) being able to be powered by clean energy such as solar energy or waste heat, (ii) Suitable for low humidity and temperature climate conditions, (iii) The rate of energy consumption rise when the humidity rate declines is much less compared to other technologies (Ejeian and Wang 2021). Consequently, this method can be economically feasible and environmentally friendly. The simple working principle of the sorption-based water harvesting method is illustrated in Fig. 8.4.

In the sorption-based systems, the adsorption process increases the partial pressure of water vapor during an open and intermittent cycle. As illustrated in Fig. 8.4, the adsorbent material adsorbs the humidity when exposed to open air because of its hydrophilicity. Then, the regeneration process starts by closing the system and heating the adsorbent material to increase its temperature leading to the desorption of the water vapor. Finally, In the condenser, water droplets are formed when water vapor comes in contact with a cold surface at a temperature below the dew point of the inside environment. These water droplets can be collected in a collection tank due to gravity.

Researchers investigated and developed numerous adsorbent materials to increase water yield in different working environments and climate conditions (Gordeeva et al. 2021; Chen et al. 2021). Despite the apparent significance of adsorbent materials, the performance of the sorption-based AWH system is affected by other factors such as the heat and mass transfer inside the system.

In the following, a brief overview and classification of the primary developed adsorbent materials for AWH systems in the literature are presented. Furthermore, a review of the different developed sorption-based AWH systems has been summarized.

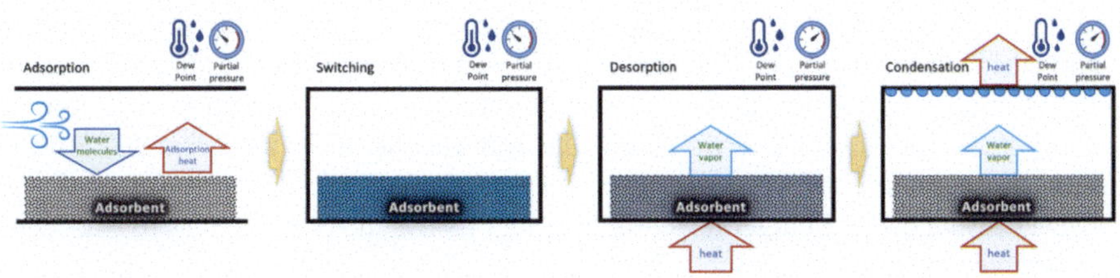

Fig. 8.4 Illustration of the working principle of the sorption-based AWH (Ejeian and Wang 2021)

Adsorbent Materials

Adsorbent materials are the critical element in any sorption-based AWH system. Numerous adsorbents have been developed over the years, and research is still ongoing to create new ones in order to improve the performance of water harvesters in addition to the other applications as well. The main preferred features of an ideal adsorbent material are (Gordeeva et al. 2021; Rocky et al. 2021):

1. Large surface area
2. Large pore volume
3. Low specific heat
4. Low regeneration temperature
5. Good packing density
6. Available and affordable
7. Large adsorption capacity
8. Uniform pore size distribution.

Needless to say, no single adsorbent material has all of these properties together. However, for every application, researchers try to develop adsorbents with the specific properties that maximize system performance.

The developed adsorbents can be classified into physical, polymeric, chemical, and composite adsorbents. In the following, an overview of the recent developments in each category has been summarized.

Physical Adsorbent

The prominent examples of physical adsorbent materials that have been investigated for AWH systems along with other applications such as cooling and desalination are Silica gel, Zeolite, and Activated carbon.

Silica Gel
It is a mesoporous material (with a Pore size volume between 2 and 50 nm) that is incompletely desiccated form of the polymeric colloidal silicic acid. Silica gel has a chemical formula written as $SiO_2.nH_2O$. The main favorite features that contribute to the intensive investigation and wide adoption of silica gel in numerous commercial products and research articles are the low regeneration temperature, affordability and commercial availability, good adsorption kinetics, and its non-toxicity. However, some disadvantageous features hinder its use in AWH such as low adsorption capacity, mainly when used in low humidity climate conditions and unfavorable thermal properties. Recent progress in research on silica gel can be found in Gado et al. (2022), Rocky et al. (2021), Askalany et al. (2013), Wang et al. (2009).

Zeolite
It belongs to the alumina silicate minerals family that its structure has a cage shape and is linked with six groups of pores that can adsorb large amounts of different molecules. So far, there are about 50 species of Zeolite found in nature (Rocky et al.

2021). Moreover, numerous synthetic zeolites are available such as NaY, ZSM-5, NaX, AQSOA FAM-Z01, AQSOA FAM-Z02, etc. However, the main disadvantages of zeolite materials are their low adsorption capacity, high required regeneration temperature, and poor thermal properties.

A lot of research has been performed over the years to overcome the shortcomings of these classical adsorbents and improve their performance and characteristics, as will be mentioned later in the chapter.

Polymeric Adsorbents

This section will cover two main polymeric adsorbents: the Metal–Organic Frameworks (MOFs) and the Hydrogels.

MOFs

MOF is considered a new class of coordination polymer comprising metal ions or clusters of them coordinated with organic ligands to form single to multiple dimensional porous crystalline structures. Over the last few decades, a lot of research has been performed on MOF materials to improve their physiochemical properties (Gordeeva et al. 2021; Rocky et al. 2021). The most exciting features of MOFs are their vast surface area, unique adsorption isotherms, structural diversity, high porosity, low density, low enthalpy of adsorption, and inclusion of guest molecules with various interactions. The ability to adjust and optimize the MOF adsorbent to the required application by adjusting the hydrophilicity, geometry, and pore size, makes it very favorable for AWH application. Nevertheless, no MOF adsorbent has been developed specifically for AWH so far. Most of the ongoing research has focused on existing MOFs (Ejeian and Wang 2021).

MOFs have some disadvantages, such as the stability of water adsorption material as the hydrolysis reaction can destroy metal–ligand bonds. Thus, strong coordination bonds are required for water-stable MOFs (Wang et al. 2016b; Furukawa et al. 2014). Furthermore, MOFs can suffer from the holes uniformity that might negatively affect macropores creation for mass transfer. Also, MOFs are still expensive and almost not available commercially, which limit their wide adoption (Ejeian and Wang 2021; Rocky et al. 2021). There is still a toxicity problem encountered with MOFs due to the presence of metal ions and solvent residues used in the synthesis process (Kumar et al. 2019). Thus, more investigations are still needed into the toxicity of MOF adsorbents. Recent work on the development of MOF for different applications, including water harvesting, can be found in Zhou et al. (2020), Hanikel et al. (2020), Liu et al. (2020).

Some MOFs show their potential in capturing carbon in the air and harvesting water from air and by adding aliphatic diamines onto the open metal sites in that way leveraging favorable acid–base chemistry between the amines and CO_2, more details can be found in Dods et al. (2022).

Hydrogels

Hydrogels can be found in various applications in daily life, such as tissue engineering to personal hygiene products. Their main advantage is the ability to absorb water from the liquid phase in large volumes. Their main interesting features that motivate researchers to consider it for AWH application are high adsorption capacity and density, operability in a wide range of relative humidity, and cyclic stability (Furukawa et al. 2014; Trapani et al. 2016). However, the main problem is the lack of porosity, which can be overcome by different methods such as controlled freezing, the use of reduced graphene oxide, and electrospinning (Kim et al. 2019; Savina et al. 2016). Moreover, relatively slow adsorption kinetics and swelling are two main drawbacks of Hydrogels (Zhao et al. 2019). Many researchers recently investigated how to improve the adsorption kinetics and water retention by adding metal bonds to polymer chains (Entezari et al. 2020; Li et al. 2018a; Kallenberger and Fröba 2018) and by using its ability to host hydroscope material such as LiCl and $CaCl_2$ (Entezari et al. 2020; Li et al. 2018a; Kallenberger and Fröba 2018; Wang et al. 2021b). Recent progress on hydrogel material development can be found in Zhao et al. (2019), Zhang et al. (2022b).

Chemical Adsorbents

For the chemical adsorbents used in AWH systems, Hygroscopic salts are the prominent example. These salts can be used in AWH application in a soluble or anhydrous form (Li et al. 2018b). Some salts can sorb water until the salt crystals are fully dissolved in it due to their high-water sorption capacity. Some salts, such as $MgSO_4$ and $CuCl_2$, have high Deliquescence Relative Humidity DRH (Represent RH at which salt transforms from the solid phase into saturated solution), which increases salt stability in the solid phase when used as adsorbent (Ejeian and Wang 2021; Li et al. 2018b). Other salt examples, such as LiCl and $CaCl_2$ (Deliquescence Salts), are characterized by high sorption capacity and the ability to sorb humidity in the solution phase by decreasing the concentration, allowing continuous water harvesting (Qi et al. 2019; Wang et al. 2019). Other merits of Hygroscopic salts are the availability and the cheap cost, besides their linear isotherms, which enables them to be used in a wide range of relative humidity climate conditions.

On the other hand, salts have some disadvantages that yet need to be solved, such as the possibility of contamination, causing corrosion to sorption devices, and limited exposure area for the case of liquid sorbents. Furthermore, high desorption enthalpy and agglomeration are two main problems with salts (Salehi et al. 2020).

Composite Adsorbent

Due to the main shortcomings of the discussed adsorbents so far, many researchers investigated developing composite adsorbents from different adsorbents to find the optimal mixture of the required properties for the application. Generally, researchers try to improve the thermal properties of the adsorbent, packing density, and volumetric and gravimetric sorption capacity depending on the application and the required properties. This can be done by the addition of Thermal Conductivity Enhancer (TCE), binders, and mixing different adsorbents together. Composite adsorbents are usually synthesized by physical mixing or chemically (Rocky et al. 2021).

For AWH applications, researchers investigated different composite adsorbents to enhance the performance of the AWH device. Hygroscopic salts have been used intensively to boost the adsorption capacity of different adsorbent materials that are used as host materials, which help in providing a stable porous environment, such as activated carbon felt and hollow carbon sphere (Entezari et al. 2019a; Li et al. 2020). Also, these host materials can supply the solution with fixed condition via the capillary effect, which allows the sorption process to continue from the solid and liquid phases (Ejeian et al. 2020). Other researchers try different host materials that have the capability to adsorb humidity as well, such as MOF and Silica gel, which help in increasing the operating range of relative humidity and the adsorption kinetics (Xu et al. 2020; Wang et al. 2016c).

Sorption Technologies

Developing a promising adsorbent with valuable features is important, but it is only one part of the AWH system design process. Other important factors that have to be taken into account when designing a water harvester are the size of the AWH device, cost, rate of energy consumption, and cycle time. Furthermore, other aspects that every designer of AWH device should consider are:

I Number of cycles per day.
II The number of sorption beds, which will affect water yield per cycle.
III Active or passive design or a hybrid device.

It is worth mentioning that the energy supply to the AWH system and whether it is active or passive, is estimated based on these main processes:

(i) Heating adsorbent material for the regeneration process.
(ii) Fresh air supply to the system for the water adsorption process.
(iii) Supply of the desorbed water vapor to the condenser and its condensation process.

Usually, passive devices have one cycle per day, in which the adsorption process occurs during the night after sunset, while the desorption process occurs during the day by solar energy after sunrise. Of course, the more passive processes an AWH device has, the less energy consumption and more economically it will be. However, this might be at the expense of the water yield of the device.

Sorption-based AWH devices' operation and water production are intermittent in nature and cannot operate continuously. However, some researchers tried to develop quasi-continuous operation by increasing the number of sorption beds with a phase difference. Numerous researchers investigated Sorption-based AWH devices based on a 24-h cycle, explained previously, mainly because it is best suited for solar

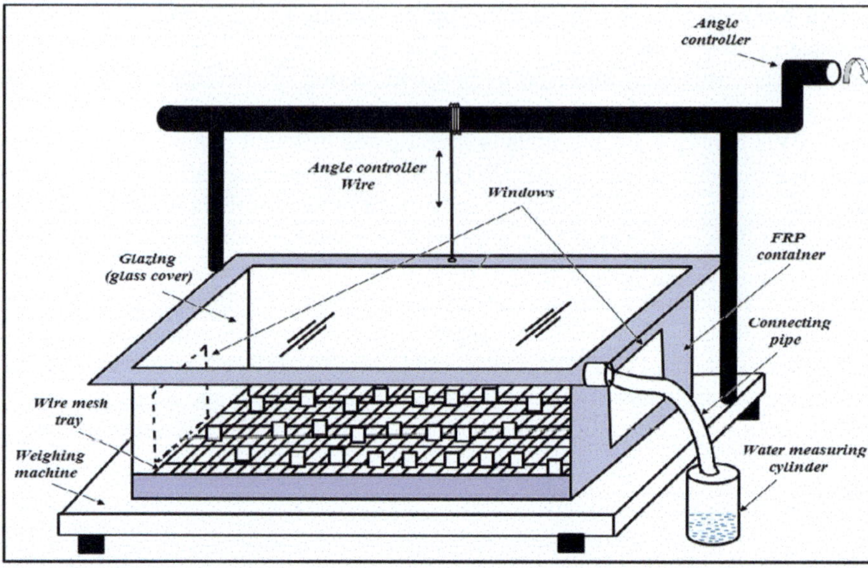

Fig. 8.5 Glass box AWH prototype developed by Kumar and Yadav (2015)

energy. However, others developed devices and prototypes that operate on several daily adsorption/desorption cycles.

Single-Cycle AWH Devices

Many researchers investigated different kinds of adsorbent materials in the one-cycle AWH device powered by solar energy. Different shapes and designs were introduced to improve the system performance, such as glass box or pyramid-shaped devices. Kumar and Yadav (2015) investigated a glass box AWH device in which the silica gel and saw wood/$CaCl_2$ are used as adsorbents. The illustration of the developed system is shown in Fig. 8.5. The glass cover is tilted at 30° and serves as a condenser for the regenerated water vapor. The water yield of the device was about 200 and 180 mL/kg per day for the two investigated adsorbents, respectively. Gandhidasan and Abualhamayel (2010) examined a glass box AWH system with a fan for air circulation, a pump for liquid $CaCl_2$ solution movement, and the upper glass was tilted by about 26.5°. The results under Saudi Arabia climate showed that the water yield per day is about 1.15 L/m^2. More similar investigations of glass box AWH devices with tilted upper glass condenser can be found in Ji et al. (2007), Kabeel (2006).

Another design of AWH device with a glass box, but the condenser was located below the adsorbent material unlike the aforementioned devices, was also investigated. This design avoids the creation of water droplets above the adsorbent material,

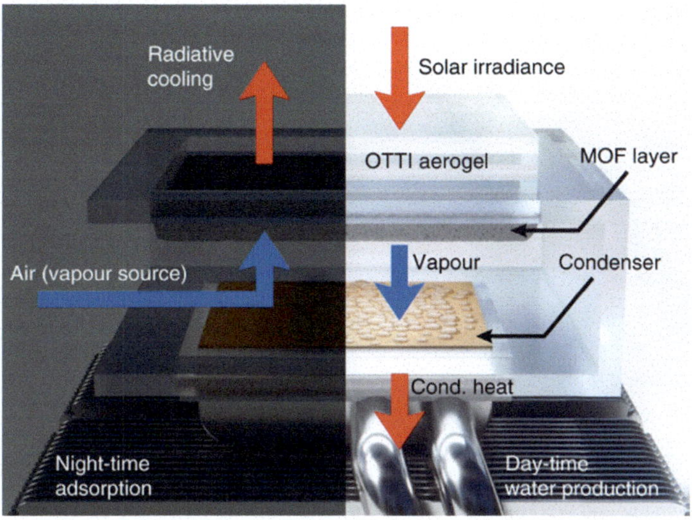

Fig. 8.6 Working principle of the AWH device developed by Kim et al. (2018)

which can negatively affect the penetration of the solar radiation into the box in addition to creating convection heat transfer flow in the device, which may lead to more thermal losses despite improving mass transfer. In these designs, the water vapor is transferred to the condenser below through its diffusivity in the inside air. However, it is worth mentioning that for these small-scale devices, the water vapor transfer is not a significant problem, but for a large-scale system, there has to be an active or passive method for mass transfer. Sleiti et al. (2021) investigated silica gel adsorbent with this design, while Kim et al. (2018) and Xu and Yaghi (2020) investigated MOFs. The basic working principle is illustrated in Fig. 8.6.

LaPotin et al. (2020) investigated a glass box AWH device in which two layers of AQSOA-Z01 layers were used to adsorb water vapor during the night. However, for the regeneration process, solar radiation desorbs the upper adsorbent layer, and the released water vapor condenses on a copper surface below the upper layer. However, this copper surface is in contact with the lower adsorbent layer to heat and regenerate it, which will lead to the recovery of the water vapor condensation latent heat. The schematic and the working principle of this device are illustrated in Fig. 8.7. The results showed a water yield of about 0.77 L/m^2 per day, which is considered 18% more compared to the single-layer configuration.

Other researchers investigated pyramid or prism-shaped glass AWH devices. The main reason is that this design allows for using multiple shelves of adsorbent materials, which help increase water yield from the AWH device without increasing the required area, as shown in Fig. 8.8. More details about this design can be found in Kabeel (2007), William et al. (2015), Gad et al. (2001).

Some researchers employ concentrating solar reflectors in their AWH design to increase the temperature of the adsorbent materials to the required level, as can be

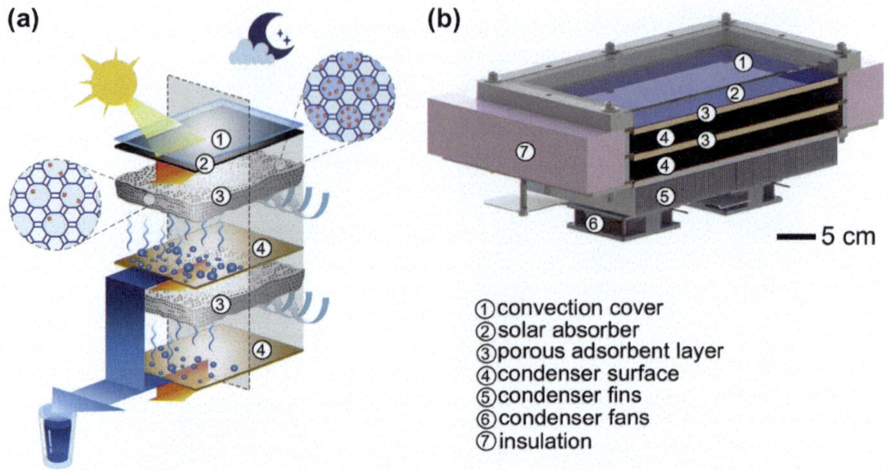

Fig. 8.7 Schematic and working principle of the dual-stage AWH device developed by LaPotin et al. (2020)

No.	Description
1,2,3	Temperature measuring points on each fiberglass side.
4	Thermocouple Selector switch position.
5	Roof surface temperature measuring point.
6	Top bed temperature measuring points.
7	Middle bed temperature measuring points.
8	Bottom bed temperature measuring points.
9	Collecting surface for condensed droplets
10	Collected water tank

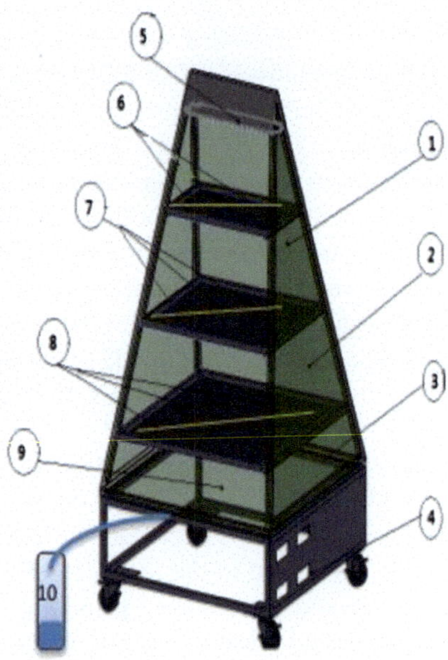

Fig. 8.8 Schematic of the glass prism-shaped AWH device developed by William et al. (2015)

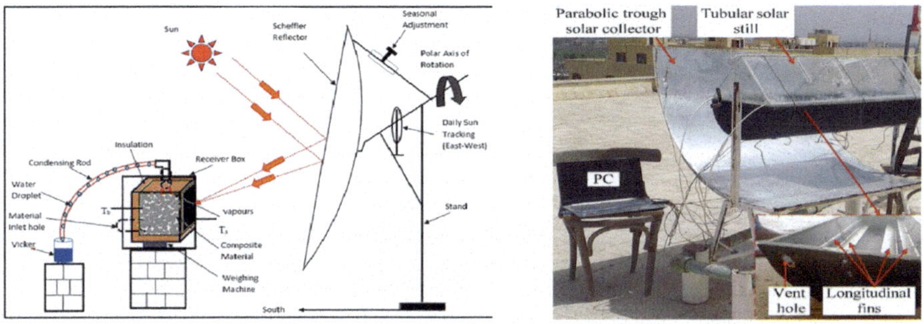

(a) **(b)**

Fig. 8.9 Some examples of AWH devices with solar concentrators; **a** Srivastava and Yadav (2018), **b** Essa et al. (2020)

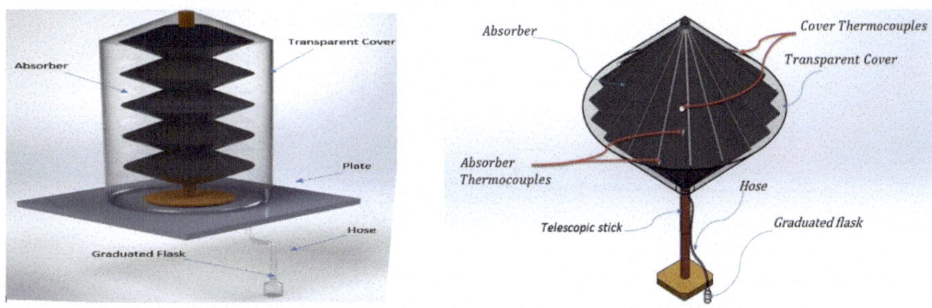

(a) **(b)**

Fig. 8.10 Some examples of the developed portable AWH devices; **a** Fathy et al. (2020a), **b** Talaat et al. (2018a)

seen in these studies (Srivastava and Yadav 2018; Essa et al. 2020; Elashmawy and Alshammari 2020; Wang et al. 2017). Some examples of these AWH designs are shown in Fig. 8.9.

Researchers investigated some AWH devices that are portable and can be easily relocated. These devices are proven to be reasonable solutions for disaster relief or camping when access to clean water is limited. Figure 8.10 illustrates some examples of portable AWH devices.

Wang et al. (2021c) investigated a different monocycle AWH prototype than the aforementioned devices. The device is an air-cooled AWH one, as shown in Fig. 8.11. The water yield of the device was about 7.7 kg_{water} per 21 $kg_{adsorbent}$ at 31° and 63% RH climate conditions.

Continuous and Multicycle AWH Devices

Continuous AWH devices should work in a similar fashion as adsorption desalination systems. However, there are no reported continuous sorption-based AWH devices

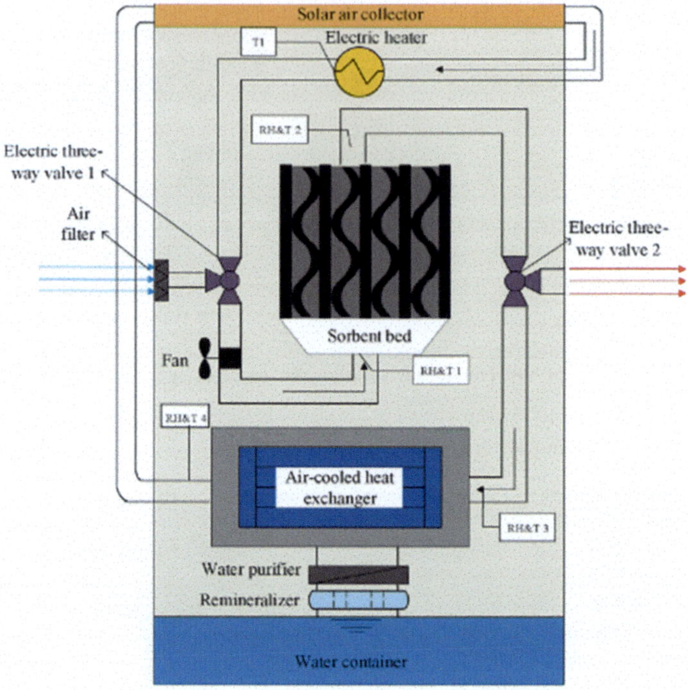

Fig. 8.11 Schematic of the single-cycle air-cooled AWH device developed by Wang et al. (2021c)

in the literature except for one recent publication by Wang et al. (2022a). In their study, an active continuous sorption-based AWH prototype, using two sorption beds made of LiCl in silica sol modified activated carbon felt adsorbent material, has been investigated. The schematic of the prototype is shown in Fig. 8.12. The results showed that the device at a regeneration temperature of 90 °C has a daily water yield of 8.3 kg_{water}/day.

However, most adsorbent materials are developed to be used in a continuous operation manner. Most published work focused on investigating different adsorbents in a single-bed configuration and multicycle adsorption operation. Li et al. (2020) proposed a new AWH device that can operate continuously, as illustrated in Fig. 8.13. The operation concept of this device depends on loading the adsorbent material, 12.6 g of a nano-carbon hollow capsule with LiCl, into a rotary cylinder, which can adsorb water vapor from atmospheric air at the bottom part for about 3 h while exposed to solar radiation at the upper part to be regenerated and release the water vapor for about 1 h as shown in Fig. 8.13. The reported water yield of this device was about 1.6 $kg_{water}/kg_{adsorbent}$ when operating on three cycles per day under 1 kW/m^2 solar radiation.

Wang et al. (2022b) investigated a dual-cycle AWH device depicted in Fig. 8.14. The used adsorbent is a composite one made from activated carbon fiber felt and an

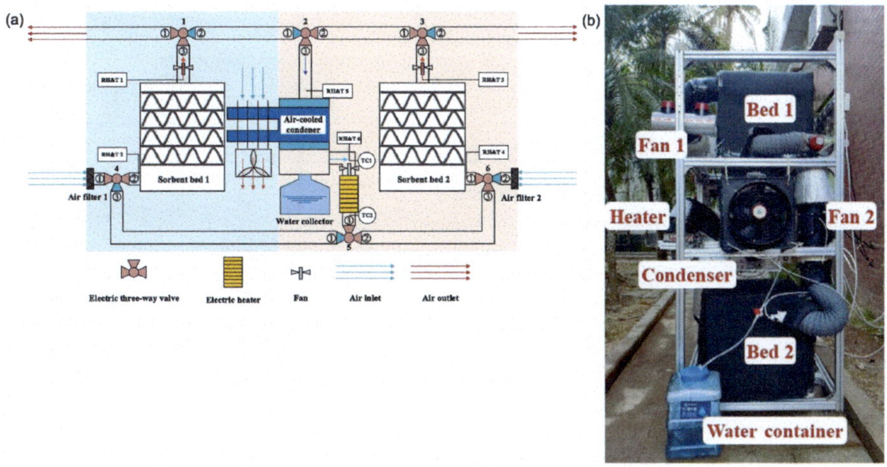

Fig. 8.12 Schematic of the continuous sorption-based AWH developed by Wang et al. (2022a)

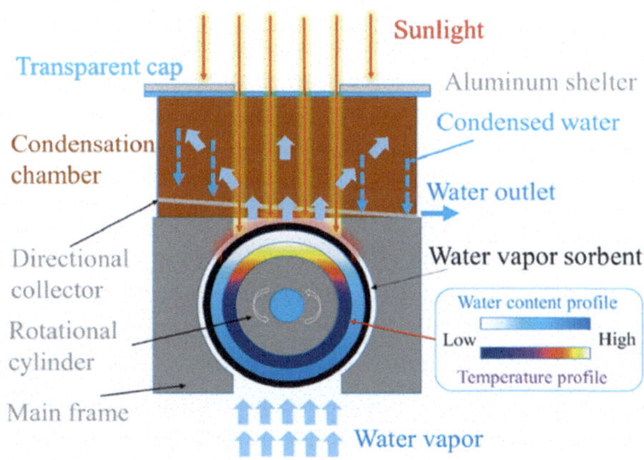

Fig. 8.13 Schematic of the AWH device developed by Li et al. (2020)

impregnated LiCl salt. The results of this study showed that the dual-cycle operation has a water yield of 0.42 kg/kg in comparison with 0.39 kg/kg for one cycle operation.

Xu et al. (2021) developed a multicycle AWH device operates on eight daily cycles and achieved a water yield of 2.12 L/kg adsorbent. Liu et al. (2021) developed an AWH prototype operating on ten cycles per day using composite LiCl/activated carbon felt adsorbent. The reported water yield is about 1.41 L/kg$_{adsorbent}$ under climate conditions of 25 °C and 75% RH.

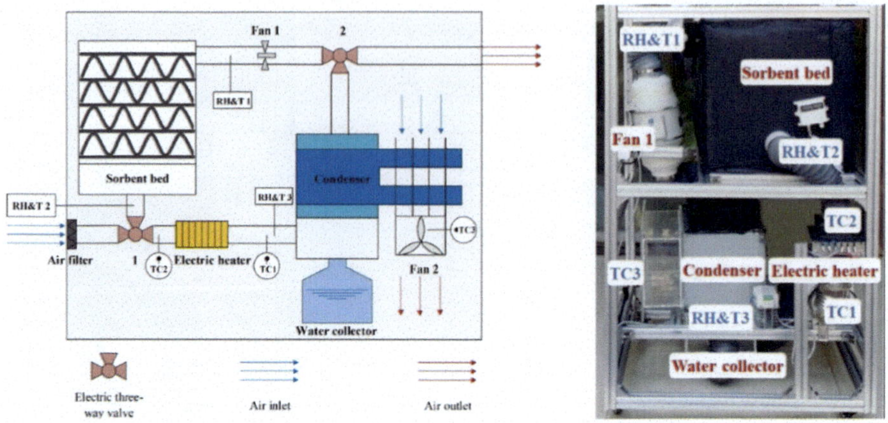

Fig. 8.14 Schematic of the AWH device developed by Wang et al. (2022b)

Possible Applications of Sorption-Based AWH Technologies

Water is generally needed for different purposes such as residential, industrial, and agricultural. Sorption-based AWH technologies can supply water for all these applications. However, each application has its own water requirements and operation restrictions regarding water quality and demand, inlet air condition, and continuous or intermittent water production. Water quality can be affected by the choice of adsorbent material, the regeneration temperature, and even the mass transfer mechanism (Ejeian and Wang 2021). Water needed for Agricultural purposes is known for not needing high-quality parameters, unlike water needed for drinking. However, some water impurities can be safe for human use but can harm some types of plants. One typical example is the use of LiCl as an adsorbent, which would increase the lithium percentage in the water, which can be safe for human use but harmful to some plants (Shahzad et al. 2016; Schrauzer 2013). Furthermore, increasing regeneration temperature can lead to an increased amount of adsorbent material traces in the harvested water (Ejeian et al. 2020). Using forced convection over passive air circulation has been proven to increase the traces of adsorbent material ions in water as in Entezari et al. (2019a), Wang et al. (2017).

For drinking purposes, a passive one-cycle sorption AWH device can barely supply the required water for drinking but not all residential water requirements at the level recommended by the world health organization of 20 L/day per person (Reed et al. 2013). Even active discontinuous AWH devices can supply about 0.22–1.05 $L/Kg_{adsorbent}$ depending on the location and climate conditions (Wang et al. 2018b). Thus, if the sorption AWH device is the only option for supplying drinking water, the device designer needs to ensure that the device can produce enough daily drinking water for residents regardless of the climate conditions parameters in terms of relative humidity and temperature throughout the year. Also, if solar energy is the thermal energy source required for the regeneration process, the AWH device design needs to consider the solar radiation throughout the year in the deployment location.

In Industrial applications, such as the steel industry, large amounts of water are used to cool down the hot molten steel, which opens an ample opportunity for recovery and recycling of this evaporated water. Furthermore, no high-quality water is needed and also waste heat is already available in such factories, which will enhance the economics of the AWH significantly. This water recovery and recycling method can be very beneficial if used to reduce the water consumption of greenhouse agriculture, as 3.3–9.5 tons of water are estimated to be released via ventilation in the semi-arid climate conditions (Entezari et al. 2019b). It is worth noting that inlet air conditions for these applications are more consistent and stable compared to atmospheric air, which will improve water productivity.

8.2.3.2 Direct Condensation (Active Cooling) Water Harvesting

The direct condensation harvesting approach represents active methods that depend on the condensation of dew water on cold surface at the expense of energy in form of electricity (Khalil et al. 2016). Compressors or vacuum pumps that run on electricity are needed for these systems, and the amount of water that can be retrieved depends on how much energy is used. Vapor compression cycle, thermoelectric generator or sorption chillers are the main technologies used for direct condensation water harvesting systems.

Using Vapor Compression Cycle (VCC)

Electricity is needed to power the VCC to produce cold sources. Cooling source temperature must be lower than the dew point temperature of the air to condense the water vapor. Two essential parameters can be a good indication for the Performance of the condensing system, first one is water harvested rate (WHR) in the unit of kg/h Eq. 8.1, and the second parameter is unit power consumption (UPC) in the unit of kWh/kg Eq. 8.2. If the WHR is high and UPC is low, the performance of system is satisfying.

$$WHR = \frac{M_w}{hr} \tag{8.1}$$

$$UPC = \frac{P}{M_w} \tag{8.2}$$

Numerous researchers investigated AWH systems using VC active systems. Zolfagharkhani et al. (2018) developed a model of gas compression refrigeration cycle to produce water from air assuming different climate conditions achieving excellent results in hot and humid area with an air velocity within the range of 2–5 m/s. A case study in south of Iran was proposed producing 22–26 L of water per day and UPC was between 0.22 and 0.3 KWh/kg.

Al-Farayedhi et al. (2014) investigated the condensate from split air conditioner (1.5 ton of refrigeration) located in Saudi Arabia and Dhahran, the results show that the maximum amount of water was in August (99.1 kg/day) when humidity is high. The daily and hourly condensate water from June to September were recorded and matched the analytical model with a maximum variation of 11.6%. Magrini et al. (2015a) discussed the opportunity of establish an integrated heat and ventilation air conditioning system (HVAC) with water production in the region of Arab Emirates coast, also a comparison between two HVAC systems is presented, the first one was an integrated system to produce water and the second was a typical system, the results revealed a reduction in the cost of water by 19% and in case of using filtration and treatment the cost reduced by 7% (Magrini et al. 2015b).

Habeebullah (2009) analyzed the performance of using evaporative coils for water extraction and predicted the amount of extracted water in August which was 16.97 kg/m^2 h and in February was 14.4 kg/m^2 h. The speed of air was optimized at 2.25 m/s because at higher velocities it is noticed that the extracted water decreased because of the insufficient thermal capacity of evaporator and at lower speeds the extracted water also decreases because of the formation of ice on evaporator surface. Using air conditioning systems and humidification-dehumidification in water desalination field was an interesting topic for scientific research (Nada et al. 2015; Elattar et al. 2016). Abu El-Maaty et al. (2021) presented an experimental and theoretical investigation of a fog desalination system as shown in Fig. 8.15, saline water is heated at an evacuated tube solar water heater (ETSWH) and pumped to solar powered heater where the saline mist is heated and evaporated towards the condenser which is cooled by refrigerator. The maximum production of fresh water was 5.83 kg/m^2. According to latest reviews (Tu and Hwang 2020) condensation technologies using VCC has UPC in the range of 0.18–2.08 KWh/kg and WHR in the range of 0.13–4.20 kg/h.

Using TEC (Thermoelectric Cooling Devices)

TEC, which are gaining popularity in academia due to their capacity to convert waste heat into electrical energy, have not yet achieved significant commercial success due to their still-relatively low efficiency. Yan et al. (2021) described a photothermally produced thermoelectric effect-based approach for dew harvesting on superhydrophobic surfaces fabricated on lithium tantalite ($LiTaO_3$). The TECs work on refrigeration at the cold end and the heat is dissipated from the hot end. The semiconductor made of material whose temperature ranges from − 130 to 90 °C (Zhang et al. 2010), is effective properly for water harvesting from air. Because the volume of semiconductor chips is smaller, water harvesting devices can be smaller and more portability than VC systems like that operated by Zhang et al. (2010) and Jardi et al. (2012a). Ibrahim et al. (2016) conducted numerical and experimental research on their own solar PV-based thermoelectric device in their paper. This evaluated its

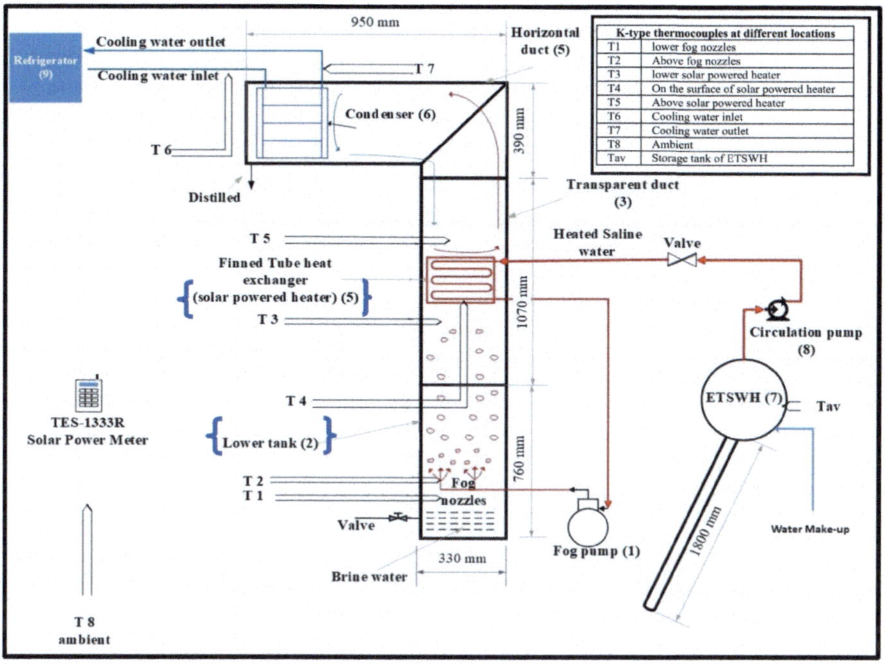

Fig. 8.15 Layout of fog desalination system (Abu El-Maaty et al. 2021)

performance and investigated promising potential for manufacturing drinking water from air using their technique. Zhang et al. (2010) performed a feasibility analysis of a TEC powered device for military deployment in remote locations with abundant solar radiation but limited water resources. Solar panels were used to generate electricity, which could be stored in batteries and used to power a TEC device for dew water harvesting. Numerous studies looked into whether employing TEC devices in the dehumidification process could cool ambient air and produce clean, safe, and constantly replenished freshwater from the moisture in the air with good quality and quantity (Tu and Hwang 2020; Milani et al. 2014). Due to their small size, light weight, ease of use, and silent operation, TEC usage are expected to grow in constructing sustainable dehumidification systems in the future. Other interesting studies using this technology can be found in references (Tu and Hwang 2020; Ibrahim et al. 2016; Milani et al. 2014; Liu et al. 2017; Jradi et al. 2012b; Eslami et al. 2018). According to latest reviews (Tu and Hwang 2020) condensation technologies using TEG has UPC in the range of 0.39–5.21 KWh/kg and WHR is lower than 1.41 kg/h.

8.2.4 Economic Analysis of AWH Systems

In this section, a cost comparison from the literature between different AWH technologies is presented. Talaat et al. (2018b) and Fathy et al. (2020b) estimated cost of producing 1 kg of drinking water from a solar-powered sorption-based AWH system as 0.062 $/kg and 0.086 $/kg respectively, they followed the same path for estimating the cost of water production using the equation below:

$$\mathrm{CPK} = \frac{\mathrm{AC}}{\mathrm{M}} \tag{8.3}$$

where CPK is the cost of 1 kg of water, AC is the annual cost including the annual cost of operational maintenance, annual salvage and fixed annual cost, M is the total production of water. The applied economic model is discussed in detail in Kabeel et al. (2010).

In The University of Texas at Austin, Wikramanayake et al. (2017) succeeded in establishing Landfill gas-powered dehumidification of air using vapor compression refrigeration cycle, which is used to get chilled cold water passing in condenser where moisture in air is condensed on the outer surface of condenser. The working technology is reducing emissions of methane by 65% and meeting 12–26% and 34% of water requirements of Kern County and Barnett oilfields. The cost analysis depended on the Net Present Value (NPV) to estimate the time-adjusted returns from the project and the Pay Back Period (PBP) to get the time required to recover the investment. Applying the equation below after estimating the water income (I), the capital expenditure (C_0), the maintenance cost (M), the depreciation (D), the trucking cost (T) and the number of years (n).

$$\mathrm{NPV} = \sum_{\mathrm{j}=0}^{\mathrm{n}} \frac{I_{t=j}}{(1+r)^j} - \left[C_0 + \sum_{j=1}^{n} \frac{\left(M_{t=j} + D_{t=j} + T_{t=j}\right)}{(1+r)^j} \right] \tag{8.4}$$

The payback period can be obtained from Eq. (8.3) for the time when NPV = 0. In Fig. 8.16 an exact price for litre of water production with a 6-year payback period is optimized, it is requiring 8.5 cents/L and 6 cents/L in the Kern County and Barnett, respectively. The results are better in the Barnett because it has more favourable weather conditions and larger landfills than in the Kern County. Table 8.1 illustrates some results from economic analysis performed on different AWH systems in the literature.

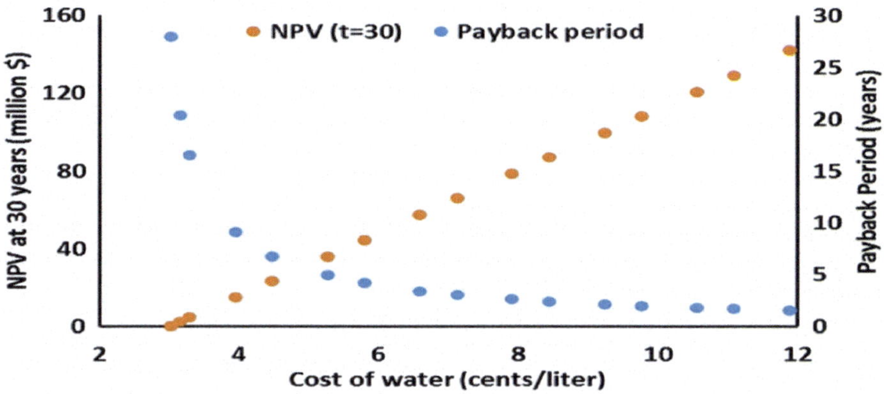

a)

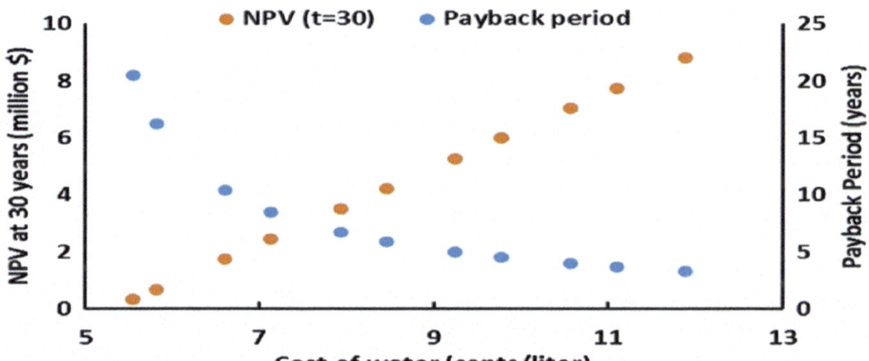

b)

Fig. 8.16 Payback period and Net present value for LFG-powered AWH projects in **a** Barnett Shale, Texas and **b** Kern County, California (Wikramanayake et al. 2017)

Table 8.1 Comparison between previous cost analysis for water extraction from air

Reference	AWH technology	Water cost
Siegel and Conser (2021)	Desiccant based AWH system	0.0065 $/kg
Deng et al. (2021)	Hygroscopic sorbent	0.84 $/kg
Talaat et al. (2018a)	Desiccant based AWH system	0.062 $/kg
Fathy et al. (2020a)	Desiccant based AWH system	0.086 $/kg
Wikramanayake et al. (2017)	VC-based AWH system	0.06–0.085 $/kg
Patel et al. (2020)	VC-based AWH system	0.038–0.24 $/kg depending on climate condition

References

Abu El-Maaty AE, Awad MM, Sultan GI, Hamed AM (2021) Performance study of fog desalination system coupled with evacuated tube solar collector. Desalination 504:114960. https://doi.org/10.1016/j.desal.2021.114960

Al-Farayedhi AA, Ibrahim NI, Gandhidasan P (2014) Condensate as a water source from vapor compression systems in hot and humid regions. Desalination 349:60–67. https://doi.org/10.1016/J.DESAL.2014.05.002

Alsaman AS, Askalany AA, Harby K, Ahmed MS (2016) A state of the art of hybrid adsorption desalination–cooling systems. Renew Sustain Energy Rev 58:692–703. https://doi.org/10.1016/J.RSER.2015.12.266

Askalany AA, Salem M, Ismael IM, Ali AHH, Morsy MG, Saha BB (2013) An overview on adsorption pairs for cooling. Renew Sustain Energy Rev 19:565–572. https://doi.org/10.1016/j.rser.2012.11.037

Alsaman AS, Hassan AA, Ali ES, Mohammed RH, Zohir AE, Farid AM, ... Askalany AA (2022) Hybrid solar-driven desalination/cooling systems: current situation and future trend. Energies 15(21):8099. https://doi.org/10.3390/en15218099

Azeem M, Guérin A, Dumais T, Caminos L, Goldstein RE, Pesci AI, de Dios Rivera J, Torres MJ, Wiener J, Campos JL, Dumais J (2020) Optimal design of multilayer fog collectors. ACS Appl Mater Interfaces 12:7736–7743. https://doi.org/10.1021/acsami.9b19727

Bilal M, Sultan M, Morosuk T, Den W, Sajjad U, Aslam MMA, Shahzad MW, Farooq M (2022) Adsorption-based atmospheric water harvesting: a review of adsorbents and systems. Int Commun Heat Mass Transfer 133:105961. https://doi.org/10.1016/J.ICHEATMASSTRANSFER.2022.105961

Cao M, Ju J, Li K, Dou S, Liu K, Jiang L (2014) Fog collection: facile and large-scale fabrication of a cactus-inspired continuous fog collector. Adv Funct Mater 24(21):3234. https://doi.org/10.1002/adfm.201470138

Cengel YA, Boles MA (n.d.) Thermodynamics: an engineering approach

Chen D, Li J, Zhao J, Guo J, Zhang S, Sherazi TA, Ambreen, Li S (2018a) Bioinspired superhydrophilic-hydrophobic integrated surface with conical pattern-shape for self-driven fog collection. J Colloid Interface Sci 530:274–281. https://doi.org/10.1016/j.jcis.2018.06.081

Chen H, Ran T, Gan Y, Zhou J, Zhang Y, Zhang L, Zhang D, Jiang L (2018b) Ultrafast water harvesting and transport in hierarchical microchannels. Nat Mater 17:935–942

Chen Z, Song S, Ma B, Li Y, Shao Y, Shi J, Liu M, Jin H, Jing D (2021) Recent progress on sorption/desorption-based atmospheric water harvesting powered by solar energy. Sol Energy Mater Sol Cells 230:111233. https://doi.org/10.1016/J.SOLMAT.2021.111233

Davies TD (1976) Precipitation scavenging of sulphur dioxide in an industrial area. Atmos Environ (1967). 10:879–890. https://doi.org/10.1016/0004-6981(76)90143-8

Deng F, Wang C, Xiang C, Wang R (2021) Bioinspired topological design of super hygroscopic complex for cost-effective atmospheric water harvesting. Nano Energy 90:106642. https://doi.org/10.1016/j.nanoen.2021.106642

de Dios Rivera J (2011) Aerodynamic collection efficiency of fog water collectors. Atmos Res 102:335–342

Dods MN, Weston SC, Long JR (2022) Prospects for simultaneously capturing carbon dioxide and harvesting water from air. Adv Mater 34:2204277. https://doi.org/10.1002/adma.202204277

Ejeian M, Wang RZ (2021) Adsorption-based atmospheric water harvesting. Joule 5:1678–1703. https://doi.org/10.1016/J.JOULE.2021.04.005

Ejeian M, Entezari A, Wang RZ (2020) Solar powered atmospheric water harvesting with enhanced LiCl /MgSO4/ACF composite. Appl Therm Eng 176:115396. https://doi.org/10.1016/j.applthermaleng.2020.115396

Elashmawy M, Alshammari F (2020) Atmospheric water harvesting from low humid regions using tubular solar still powered by a parabolic concentrator system. J Clean Prod 256:120329. https://doi.org/10.1016/j.jclepro.2020.120329

Elattar HF, Fouda A, Nada SA (2016) Performance investigation of a novel solar hybrid air conditioning and humidification–dehumidification water desalination system. Desalination 382:28–42. https://doi.org/10.1016/j.desal.2015.12.023

Entezari A, Ejeian M, Wang RZ (2019a) Extraordinary air water harvesting performance with three phase sorption. Mater Today Energy 13:362–373. https://doi.org/10.1016/j.mtener.2019.07.001

Entezari A, Wang RZ, Zhao S, Mahdinia E, Wang JY, Tu YD, Huang DF (2019b) Sustainable agriculture for water-stressed regions by air-water-energy management. Energy 181:1121–1128. https://doi.org/10.1016/J.ENERGY.2019.06.045

Entezari A, Ejeian M, Wang R (2020) Super atmospheric water harvesting hydrogel with alginate chains modified with binary salts. ACS Mater Lett 2:471–477. https://doi.org/10.1021/acsmaterialslett.9b00315

Eslami M, Tajeddini F, Etaati N (2018) Thermal analysis and optimization of a system for water harvesting from humid air using thermoelectric coolers. Energy Convers Manag 174:417–429. https://doi.org/10.1016/j.enconman.2018.08.045

Essa FA, Elsheikh AH, Sathyamurthy R, Muthu Manokar A, Kandeal AW, Shanmugan S, Kabeel AE, Sharshir SW, Panchal H, Younes MM (2020) Extracting water content from the ambient air in a double-slope half-cylindrical basin solar still using silica gel under Egyptian conditions. Sustain Energy Technol Assess 39:100712. https://doi.org/10.1016/j.seta.2020.100712

Fathy MH, Awad MM, Zeidan ESB, Hamed AM (2020a) Solar powered foldable apparatus for extracting water from atmospheric air. Renew Energy 162:1462–1489. https://doi.org/10.1016/j.renene.2020.07.020

Fathy MH, Awad MM, Zeidan E-SB, Hamed AM (2020b) Solar powered foldable apparatus for extracting water from atmospheric air. Renew Energy 162:1462–1489. https://doi.org/10.1016/j.renene.2020b.07.020

Fernandez DM, Kleingartner J, Oliphant A, Bowman M, Torregrosa A, Weiss-Penzias PS, Zhang BJ, Sorensen D, Cohen RE, McKinley GH (2018) Fog water collection effectiveness: mesh intercomparisons. Aerosol Air Qual Res 18:270–283. https://doi.org/10.4209/aaqr.2017.01.0040

Furukawa H, Gándara F, Zhang YB, Jiang J, Queen WL, Hudson MR, Yaghi OM (2014) Water adsorption in porous metal-organic frameworks and related materials. J Am Chem Soc 136:4369–4381. https://doi.org/10.1021/JA500330A/SUPPL_FILE/JA500330A_SI_010.PDF

Gad HE, Hamed AM, El-Sharkawy II (2001) Application of a solar desiccant/collector system for water recovery from atmospheric air. Renew Energy 22:541–556. https://doi.org/10.1016/S0960-1481(00)00112-9

Gado MG, Nasser M, Hassan AA, Hassan H (2022) Adsorption-based atmospheric water harvesting powered by solar energy: comprehensive review on desiccant materials and systems. Process Saf Environ Prot 160:166–183. https://doi.org/10.1016/j.psep.2022.01.061

Gandhidasan P, Abualhamayel HI (2010) Investigation of humidity harvest as an alternative water source in the Kingdom of Saudi Arabia. Water Environ J 24:282–292. https://doi.org/10.1111/j.1747-6593.2009.00189.x

Gao Y, Wang J, Xia W, Mou X, Cai Z (2018) Reusable hydrophilic-superhydrophobic patterned weft backed woven fabric for high-efficiency water-harvesting application. ACS Sustain Chem Eng 6:7216–7220

Ghosh R, Ray TK, Ganguly R (2015) Cooling tower fog harvesting in power plants—a pilot study. Energy 89:1018–1028. https://doi.org/10.1016/j.energy.2015.06.050

Good RJ (1992) Contact angle, wetting, and adhesion: a critical review. J Adhes Sci Technol 6:1269–1302. https://doi.org/10.1163/156856192X00629

Gordeeva LG, Tu YD, Pan Q, Palash ML, Saha BB, Aristov YI, Wang RZ (2021) Metal-organic frameworks for energy conversion and water harvesting: a bridge between thermal engineering and material science. Nano Energy 84:105946. https://doi.org/10.1016/J.NANOEN.2021.105946

Greenlee LF, Lawler DF, Freeman BD, Marrot B, Moulin P (2009) Reverse osmosis desalination: water sources, technology, and today's challenges. Water Res 43:2317–2348. https://doi.org/10.1016/j.watres.2009.03.010

Guo J, Huang W, Guo Z, Liu W (2022) Design of a venation-like patterned surface with hybrid wettability for highly efficient fog harvesting. Nano Lett 22:3104–3111. https://doi.org/10.1021/acs.nanolett.2c00488

Habeebullah BA (2009) Potential use of evaporator coils for water extraction in hot and humid areas. Desalination 237:330–345. https://doi.org/10.1016/j.desal.2008.01.025

Hanikel N, Prévot MS, Yaghi OM (2020) MOF water harvesters. Nat Nanotechnol 15(5):348–355. https://doi.org/10.1038/s41565-020-0673-x

Hassan AA, Elwardany AE, Ookawara S, El-Sharkawy II (2020a) Performance investigation of integrated PVT/adsorption cooling system under the climate conditions of Middle East. Energy Rep 6:168–173. https://doi.org/10.1016/j.egyr.2020.11.096

Hassan AA, Elwardany AE, Ookawara S, Ahmed M, El-Sharkawy II (2020b) Integrated adsorption-based multigeneration systems: a critical review and future trends. Int J Refrig 116:129–145. https://doi.org/10.1016/j.ijrefrig.2020.04.001

Hassan AA, Elwardany AE, Ookawara S, El-Sharkawy II (2021) Performance investigation of a solar-powered adsorption-based trigeneration system for cooling, electricity, and domestic hot water production. Appl Therm Eng 199:117553. https://doi.org/10.1016/j.applthermaleng.2021.117553

Hassan AA, Elwardany AE, Ookawara S, Sekiguchi H, Hassan H (2022a) Performance and economic analysis of hybrid solar collectors-powered integrated adsorption/reverse osmosis multigeneration system. Int J Energy Res. https://doi.org/10.1002/ER.8512

Hassan AA, Elwardany AE, Ookawara S, Sekiguchi H, Hassan H (2022b) Energy, exergy, economic and environmental (4E) assessment of hybrid solar system powering adsorption-parallel/series ORC multigeneration system. Process Saf Environ Prot. https://doi.org/10.1016/J.PSEP.2022.06.024

Ibrahim NI, Al-Sulaiman FA, Saidur R (2016) Performance assessment of water production from solar cooling system in humid climate. Energy Convers Manag 127:647–655. https://doi.org/10.1016/j.enconman.2016.09.056

Ji JG, Wang RZ, Li LX (2007) New composite adsorbent for solar-driven fresh water production from the atmosphere. Desalination 212:176–182. https://doi.org/10.1016/j.desal.2006.10.008

Jradi M, Ghaddar N, Ghali K (2012a) Optimized operation of a solar-driven thermoelectric dehumidification system for fresh water production. Energy Power Eng 6:878–891

Jradi M, Ghaddar NK, Ghali KF (2012b) Experimental and theoretical study of an integrated thermoelectric–photovoltaic system for air dehumidification and fresh water production. Int J Energy Res 36

Kabeel AE (2006) Application of sandy bed solar collector system for water extraction from air. Int J Energy Res 30:381–394. https://doi.org/10.1002/er.1155

Kabeel AE (2007) Water production from air using multi-shelves solar glass pyramid system. Renew Energy 32:157–172. https://doi.org/10.1016/j.renene.2006.01.015

Kabeel AE, Hamed AM, El-Agouz SA (2010) Cost analysis of different solar still configurations. Energy 35:2901–2908. https://doi.org/10.1016/j.energy.2010.03.021

Kallenberger PA, Fröba M (2018) Water harvesting from air with a hygroscopic salt in a hydrogel–derived matrix. Commun Chem 1:6–11. https://doi.org/10.1038/s42004-018-0028-9

Khalil B, Adamowski J, Shabbir A, Jang C, Rojas M, Reilly K, Ozga-Zielinski B (2016) A review: dew water collection from radiative passive collectors to recent developments of active collectors. Sustain Water Resour Manag 2:71–86. https://doi.org/10.1007/S40899-015-0038-Z/FIGURES/8

Kim H, Rao SR, Kapustin EA, Zhao L, Yang S, Yaghi OM, Wang EN (2018) Adsorption-based atmospheric water harvesting device for arid climates. Nat Commun 9:1–8. https://doi.org/10.1038/s41467-018-03162-7

Kim S, Park H, Choi H (2019) Maneuvering the ordered mesoporosity of electrospun silica nanofibers for water harvesting. Microporous Mesoporous Mater 281:23–31. https://doi.org/10.1016/J.MICROMESO.2019.02.037

Kumar M, Yadav A (2015) Experimental investigation of solar powered water production from atmospheric air by using composite desiccant material "$CaCl_2$/saw wood." Desalination 367:216–222. https://doi.org/10.1016/j.desal.2015.04.009

Kumar P, Anand B, Tsang YF, Kim KH, Khullar S, Wang B (2019) Regeneration, degradation, and toxicity effect of MOFs: opportunities and challenges. Environ Res 176:108488. https://doi.org/10.1016/J.ENVRES.2019.05.019

LaPotin A, Zhong Y, Zhang L, Zhao L, Leroy A, Kim H, Rao SR, Wang EN (2020) Dual-stage atmospheric water harvesting device for scalable solar-driven water production. Joule 1–17. https://doi.org/10.1016/j.joule.2020.09.008

Li R, Shi Y, Alsaedi M, Wu M, Shi L, Wang P (2018a) Hybrid hydrogel with high water vapor harvesting capacity for deployable solar-driven atmospheric water generator. Environ Sci Technol 52:11367–11377. https://doi.org/10.1021/acs.est.8b02852

Li R, Shi Y, Shi L, Alsaedi M, Wang P (2018b) Harvesting water from air: using anhydrous salt with sunlight. Environ Sci Technol 52:5398–5406. https://doi.org/10.1021/acs.est.7b06373

Li C, Liu Y, Gao C, Li X, Xing Y, Zheng Y (2019) Fog harvesting of a bioinspired nanocone-decorated 3D fiber network. ACS Appl Mater Interfaces 11:4507–4513. https://doi.org/10.1021/acsami.8b15901

Li R, Shi Y, Wu M, Hong S, Wang P (2020) Improving atmospheric water production yield: enabling multiple water harvesting cycles with nano sorbent. Nano Energy 67:104255. https://doi.org/10.1016/j.nanoen.2019.104255

Liu S, He W, Hu D, Lv S, Chen D, Wu X, Xu F, Li S (2017) Experimental analysis of a portable atmospheric water generator by thermoelectric cooling method. Energy Procedia 142:1609–1614

Liu H, Xie W-Y, Song F, Wang X-L, Wang Y-Z (2019) Constructing hierarchically hydrophilic/superhydrophobic ZIF-8 pattern on soy protein towards a biomimetic efficient water harvesting material. Chem Eng J 369:1040–1048. https://doi.org/10.1016/j.cej.2019.03.152

Liu X, Wang X, Kapteijn F (2020) Water and metal-organic frameworks: from interaction toward utilization. Chem Rev 120:8303–8377. https://doi.org/10.1021/ACS.CHEMREV.9B00746/ASSET/IMAGES/LARGE/CR9B00746_0019.JPEG

Liu XY, Wang WW, Xie ST, Pan QW (2021) Performance characterization and application of composite adsorbent LiCl@ACFF for moisture harvesting. Sci Rep 11(1):1–10. https://doi.org/10.1038/s41598-021-93784-7

Liu X, Beysens D, Bourouina T (2022) Water harvesting from air: current passive approaches and outlook. ACS Mater Lett 4:1003–1024. https://doi.org/10.1021/acsmaterialslett.1c00850

Lu H, Shi W, Guo Y, Guan W, Lei C, Yu G (2022) Materials engineering for atmospheric water harvesting: progress and perspectives. Adv Mater 34:2110079. https://doi.org/10.1002/adma.202110079

Magrini A, Cattani L, Cartesegna M, Magnani L (2015a) Integrated systems for air conditioning and production of drinking water—preliminary considerations. Energy Procedia 75:1659–1665. https://doi.org/10.1016/j.egypro.2015.07.406

Magrini A, Cattani L, Cartesegna M, Magnani L (2015b) Production of water from the air: the environmental sustainability of air-conditioning systems through a more intelligent use of resources. the advantages of an integrated system. Energy Procedia 78:1153–1158. https://doi.org/10.1016/j.egypro.2015.11.081

Milani D, Qadir A, Vassallo A, Chiesa M, Abbas A (2014) Experimentally validated model for atmospheric water generation using a solar assisted desiccant dehumidification system. Energy Build 77:236–246. https://doi.org/10.1016/j.enbuild.2014.03.041

Moossa B, Trivedi P, Saleem H, Zaidi SJ (2022) Desalination in the GCC countries—a review. J Clean Prod 357:131717. https://doi.org/10.1016/J.JCLEPRO.2022.131717

Nada SA, Elattar HF, Fouda A (2015) Performance analysis of proposed hybrid air conditioning and humidification–dehumidification systems for energy saving and water production in hot and dry climatic regions. Energy Convers Manag 96:208–227. https://doi.org/10.1016/j.enconman.2015.02.082

Olivier J (2004) Fog-water harvesting along the West Coast of South Africa: a feasibility study. Water SA 28. https://doi.org/10.4314/wsa.v28i4.4908

Pan Z, Pitt WG, Zhang Y-M, Wu N, Tao Y, Truscott TT (2016) The upside-down water collection system of Syntrichia caninervis. Nat Plants 2

Parker AR, Lawrence CR (2001a) Water capture by a desert beetle. Nature 414:6859:33–34. https://doi.org/10.1038/35102108

Parker AR, Lawrence CR (2001b) Water capture by a desert beetle. Nature 414:33–34. https://doi.org/10.1038/35102108

Patel J, Patel K, Mudgal A, Panchal H, Sadasivuni KK (2020) Experimental investigations of atmospheric water extraction device under different climatic conditions. Sustain Energy Technol Assess 38:100677. https://doi.org/10.1016/J.SETA.2020.100677

Pei M, Huo L, Zhao X, Chen S, Li J, Peng Z, Zhang K, Zhou H, Liu P (2020) Facile construction of stable hydrophobic surface via covalent self-assembly of silane-terminated fluorinated polymer. Appl Surf Sci 507:145138. https://doi.org/10.1016/j.apsusc.2019.145138

Peng Y, He Y, Yang S, Ben S, Cao M, Li K, Liu K, Jiang L (2015) Magnetically induced fog harvesting via flexible conical arrays. Adv Funct Mater 25:5967–5971. https://doi.org/10.1002/adfm.201502745

Pimentel-Rodrigues C, Silva-Afonso A (2022) Rainwater harvesting for irrigation of tennis courts: a case study. Water (Switzerland) 14:1–10. https://doi.org/10.3390/w14050752

Qi H, Wei T, Zhao W, Zhu B, Liu G, Wang P, Lin Z, Wang X, Li X, Zhang X, Zhu J (2019) An interfacial solar-driven atmospheric water generator based on a liquid sorbent with simultaneous adsorption-desorption. Adv Mater 31:1–9. https://doi.org/10.1002/adma.201903378

Raut HK, Ranganath AS, Baji A, Wood KL (2019) Bio-inspired hierarchical topography for texture driven fog harvesting. Appl Surf Sci 465:362–368. https://doi.org/10.1016/j.apsusc.2018.09.134

Reed R, Godfrey S, Kayaga S, Reed B, Rouse J, Fisher J, Vilholth K, Odhiambo F (2013) Technical notes on drinking-water, sanitation and hygiene in emergencies

Rocky KA, Pal A, Rupam TH, Palash ML, Saha BB (2021) Recent advances of composite adsorbents for heat transformation applications. Therm Sci Eng Prog 23:100900. https://doi.org/10.1016/J.TSEP.2021.100900

Salehi AA, Ghannadi-Maragheh M, Torab-Mostaedi M, Torkaman R, Asadollahzadeh M (2020) A review on the water-energy nexus for drinking water production from humid air. Renew Sustain Energy Rev 120:109627. https://doi.org/10.1016/J.RSER.2019.109627

Savina IN, Ingavle GC, Cundy AB, Mikhalovsky SV (2016) A simple method for the production of large volume 3D macroporous hydrogels for advanced biotechnological, medical and environmental applications. Sci Rep 6(1):1–9. https://doi.org/10.1038/srep21154

Schrauzer GN (2013) Lithium: occurrence, dietary intakes, nutritional essentiality. 21:14–21. https://doi.org/10.1080/07315724.2002.10719188

Shahzad B, Tanveer M, Hassan W, Shah AN, Anjum SA, Cheema SA, Ali I (2016) Lithium toxicity in plants: reasons, mechanisms and remediation possibilities—a review. Plant Physiol Biochem 107:104–115. https://doi.org/10.1016/J.PLAPHY.2016.05.034

Shanyengana ES, Sanderson RD, Seely MK, Schemenauer RS (2003) Testing greenhouse shade nets in collection of fog for water supply. J Water Supply: Res Technol—AQUA 52:237–241. https://doi.org/10.2166/aqua.2003.0023

Sharma VP, Arora HC, Gupta RK (1983) Atmospheric pollution studies at Kanpur—suspended particulate matter. Atmos Environ (1967) 17:1307–1313. https://doi.org/10.1016/0004-6981(83)90405-5

Sharma V, Yiannacou K, Karjalainen M, Lahtonen K, Valden M, Sariola V (2019) Large-scale efficient water harvesting using bioinspired micro-patterned copper oxide nanoneedle surfaces and guided droplet transport. Nanoscale Adv

Shukla JB, Misra AK, Sundar S, Naresh R (2008) Effect of rain on removal of a gaseous pollutant and two different particulate matters from the atmosphere of a city. Math Comput Model 48:832–844. https://doi.org/10.1016/j.mcm.2007.10.016

Shukla JB, Misra AK, Naresh R, Chandra P (2010) How artificial rain can be produced? A mathematical model. Nonlinear Anal Real World Appl 11:2659–2668. https://doi.org/10.1016/j.nonrwa.2009.09.013

Siegel NP, Conser B (2021) A techno-economic analysis of solar-driven atmospheric water harvesting. J Energy Res Technol Trans ASME 143. https://doi.org/10.1115/1.4049286

Sleiti AK, Al-Khawaja H, Al-Khawaja H, Al-Ali M (2021) Harvesting water from air using adsorption material—prototype and experimental results. Sep Purif Technol 257:117921. https://doi.org/10.1016/j.seppur.2020.117921

Söz ÇK, Trosien S, Biesalski M (2020) Janus interface materials: a critical review and comparative study. ACS Mater Lett 2:336–357. https://doi.org/10.1021/acsmaterialslett.9b00489

Srivastava S, Yadav A (2018) Water generation from atmospheric air by using composite desiccant material through fixed focus concentrating solar thermal power. Sol Energy 169:302–315. https://doi.org/10.1016/j.solener.2018.03.089

Su Y, Cai S, Wu T, Li C, Huang Z, Zhang Y, Wu H, Hu K, Chen C, Li J, Hu Y, Zhu S, Wu D (2019) Smart stretchable Janus membranes with tunable collection rate for fog harvesting. Adv Mater Interfaces 6:1901465. https://doi.org/10.1002/admi.201901465

Summa S, Tarabelli L, Di Perna C (2021) Evaluation of ISO 52010–1: 2017 and proposal for an alternative calculation procedure. Sol Energy 218:262–281. https://doi.org/10.1016/j.solener.2021.02.046

Talaat MA, Awad MM, Zeidan EB, Hamed AM (2018a) Solar-powered portable apparatus for extracting water from air using desiccant solution. Renew Energy 119:662–674. https://doi.org/10.1016/j.renene.2017.12.050

Talaat MA, Awad MM, Zeidan EB, Hamed AM (2018b) Solar-powered portable apparatus for extracting water from air using desiccant solution. Renew Energy 119:662–674. https://doi.org/10.1016/j.renene.2017.12.050

Trapani F, Polyzoidis A, Loebbecke S, Piscopo CG (2016) On the general water harvesting capability of metal-organic frameworks under well-defined climatic conditions. Microporous Mesoporous Mater 230:20–24. https://doi.org/10.1016/j.micromeso.2016.04.040

Tu Y, Wang R, Zhang Y, Wang J (2018) Progress and expectation of atmospheric water harvesting. Joule 2:1452–1475. https://doi.org/10.1016/j.joule.2018.07.015

Tu R, Hwang Y (2020) Reviews of atmospheric water harvesting technologies. Energy 201:117630. https://doi.org/10.1016/j.energy.2020.117630

Wang LW, Wang RZ, Oliveira RG (2009) A review on adsorption working pairs for refrigeration. Renew Sustain Energy Rev 13:518–534. https://doi.org/10.1016/j.rser.2007.12.002

Wang B, Zhang Y, Liang W, Wang G, Guo Z, Liu W (2014) A simple route to transform normal hydrophilic cloth into a superhydrophobic–superhydrophilic hybrid surface. J Mater Chem A 2:7845–7852. https://doi.org/10.1039/C4TA00833B

Wang Y, Wang X, Lai C, Hu H, Kong Y, Fei B, Xin JH (2016a) Biomimetic water-collecting fabric with light-induced superhydrophilic bumps. ACS Appl Mater Interfaces. 8(5):2950–2960

Wang C, Liu X, Keser Demir N, Chen JP, Li K (2016b) Applications of water stable metal–organic frameworks. Chem Soc Rev 45:5107–5134. https://doi.org/10.1039/C6CS00362A

Wang JY, Wang RZ, Wang LW (2016c) Water vapor sorption performance of ACF-$CaCl_2$ and silica gel-$CaCl_2$ composite adsorbents. Appl Therm Eng 100:893–901. https://doi.org/10.1016/J.APPLTHERMALENG.2016.02.100

Wang JY, Wang RZ, Wang LW, Liu JY (2017) A high efficient semi-open system for fresh water production from atmosphere. Energy 138:542–551. https://doi.org/10.1016/j.energy.2017.07.106

Wang J-N, Liu Y-Q, Zhang Y-L, Feng J, Wang H, Yu Y-H, Sun H-B (2018a) Wearable superhydrophobic elastomer skin with switchable wettability. Adv Funct Mater 28:1800625. https://doi.org/10.1002/adfm.201800625

Wang JY, Wang RZ, Tu YD, Wang LW (2018b) Universal scalable sorption-based atmosphere water harvesting. Energy 165:387–395. https://doi.org/10.1016/J.ENERGY.2018.09.106

Wang X, Li X, Liu G, Li J, Hu X, Xu N, Zhao W, Zhu B, Zhu J (2019) An interfacial solar heating assisted liquid sorbent atmospheric water generator. Angew Chem 131:12182–12186. https://doi.org/10.1002/ange.201905229

Wang B, Zhou X, Guo Z, Liu W (2021a) Recent advances in atmosphere water harvesting: design principle, materials, devices, and applications. Nano Today 40:101283. https://doi.org/10.1016/j.nantod.2021.101283

Wang M, Sun T, Wan D, Dai M, Ling S, Wang J, Liu Y, Fang Y, Xu S, Yeo J, Yu H, Liu S, Wang Q, Li J, Yang Y, Fan Z, Chen W (2021b) Solar-powered nanostructured biopolymer hygroscopic aerogels for atmospheric water harvesting. Nano Energy 80:105569. https://doi.org/10.1016/J.NANOEN.2020.105569

Wang W, Xie S, Pan Q, Dai Y, Wang R, Ge T (2021c) Air-cooled adsorption-based device for harvesting water from island air. Renew Sustain Energy Rev 141:110802. https://doi.org/10.1016/J.RSER.2021.110802

Wang W, Yang T, Pan Q, Dai Y, Wang R, Ge T (2022a) All-day freshwater production enabled by an active continuous sorption-based atmospheric water harvesting system. Energy Convers Manag 264:115745. https://doi.org/10.1016/J.ENCONMAN.2022.115745

Wang W, Pan Q, Xing Z, Liu X, Dai Y, Wang R, Ge T (2022b) Viability of a practical multicyclic sorption-based water harvester with improved water yield. Water Res 211:118029. https://doi.org/10.1016/J.WATRES.2021.118029

Wikramanayake ED, Ozkan O, Bahadur V (2017) Landfill gas-powered atmospheric water harvesting for oilfield operations in the United States. Energy 138:647–658. https://doi.org/10.1016/j.energy.2017.07.062

William GE, Mohamed MH, Fatouh M (2015) Desiccant system for water production from humid air using solar energy. Energy 90:1707–1720. https://doi.org/10.1016/j.energy.2015.06.125

Xu W, Yaghi OM (2020) Metal-organic frameworks for water harvesting from air, anywhere, anytime. ACS Cent Sci 6:1348–1354. https://doi.org/10.1021/acscentsci.0c00678

Xu C, Feng R, Song F, Wang X-L, Wang Y-Z (2018) Desert beetle-inspired superhydrophilic/superhydrophobic patterned cellulose film with efficient water collection and antibacterial performance. ACS Sustain Chem Eng 6:14679–14684. https://doi.org/10.1021/acssuschemeng.8b03247

Xu J, Li T, Chao J, Wu S, Yan T, Li W, Cao B, Wang R (2020) Efficient solar-driven water harvesting from arid air with metal-organic frameworks modified by hygroscopic salt. Angew Chem Int Ed 59:5202–5210. https://doi.org/10.1002/ANIE.201915170

Xu J, Li T, Yan T, Wu S, Wu M, Chao J, Huo X, Wang P, Wang R (2021) Ultrahigh solar-driven atmospheric water production enabled by scalable rapid-cycling water harvester with vertically aligned nanocomposite sorbent. Energy Environ Sci 14:5979–5994. https://doi.org/10.1039/D1EE01723C

Yan W, Zhao C, Luo W, Zhang W, Li X, Liu D (2021) Optically guided pyroelectric manipulation of water droplet on a superhydrophobic surface. ACS Appl Mater Interfaces 13:23181–23190. https://doi.org/10.1021/acsami.1c03407

Yi S, Wang J, Chen Z, Liu B, Ren L, Liang L, Jiang L (2019) Fog harvesting system: cactus-inspired conical spines with oriented microbarbs for efficient fog harvesting. Adv Mater Technol 4(12):1970068

Yin K, Du H, Dong X, Wang C, Duan JA, He J (2017) A simple way to achieve bioinspired hybrid wettability surface with micro/nanopatterns for efficient fog collection. Nanoscale 9:14620–14626. https://doi.org/10.1039/C7NR05683D

Zhang T, Wan RH, Yan T, Wang L, Lin CH (2010) Development of portable air-based water collector for field operations under conditions of severe water shortage. Chin Med Equip J 31:70–72

Zhang Q, Lin G, Yin J (2018) Highly efficient fog harvesting on superhydrophobic microfibers through droplet oscillation and sweeping. Soft Matter 14:8276–8283. https://doi.org/10.1039/C8SM01688G

Zhang Y, Zhu W, Zhang C, Peoples J, Li X, Felicelli AL, Shan X, Warsinger DM, Borca-Tasciuc T, Ruan X, Li T (2022a) Atmospheric water harvesting by large-scale radiative cooling cellulose-based fabric. Nano Lett 22:2618–2626. https://doi.org/10.1021/acs.nanolett.1c04143

Zhang Z, Fu H, Li Z, Huang J, Xu Z, Lai Y, Qian X, Zhang S (2022b) Hydrogel materials for sustainable water resources harvesting and treatment: synthesis, mechanism and applications. Chem Eng J 439:135756. https://doi.org/10.1016/J.CEJ.2022.135756

Zhao F, Zhou X, Liu Y, Shi Y, Dai Y, Yu G, Zhao F, Zhou X, Shi Y, Yu G, Liu Y, Dai Y, Dai YF (2019) Super moisture-absorbent gels for all-weather atmospheric water harvesting. Adv Mater 31:1806446. https://doi.org/10.1002/ADMA.201806446

Zheng Y, Bai H, Huang Z, Tian X, Nie FQ, Zhao Y, Zhai J, Jiang L (2010) Directional water collection on wetted spider silk. Nature 463:640–643. https://doi.org/10.1038/NATURE08729

Zhou H, Zhang M, Li C, Gao C, Zheng Y (2018) Excellent fog-droplets collector via integrative Janus membrane and conical spine with micro/nanostructures. Small 14(27):e1801335

Zhou X, Lu H, Zhao F, Yu G (2020) Atmospheric water harvesting: a review of material and structural designs. ACS Mater Lett 2:671–684. https://doi.org/10.1021/ACSMATERIALSLETT.0C00130/ASSET/IMAGES/LARGE/TZ0C00130_0009.JPEG

Zhu H, Guo Z, Liu W (2016) Biomimetic water-collecting materials inspired by nature. Chem Commun 52:3863–3879. https://doi.org/10.1039/C5CC09867J

Zolfagharkhani S, Zamen M, Shahmardan MM (2018) Thermodynamic analysis and evaluation of a gas compression refrigeration cycle for fresh water production from atmospheric air. Energy Convers Manag 170:97–107. https://doi.org/10.1016/j.enconman.2018.05.016

Chapter 9
Businesses Based on Atmospheric Water Harvesting Around the World

Elvis Fosso-Kankeu, Thabo T. I. Nkambule, and Bhekie B. Mamba

9.1 Introduction

With the growing world population, water scarcity has been escalating over the years as a result of increasing industrial activities and expansion of irrigated agriculture which are the main impetus in the rising global water demand. It is expected that the already complex relationship between world development and water demands will be worsen following climate change coupled with increase in bioenergy demand. According to some expert predictions, approximately 6 billion of the world's population will be living in potential water scarce areas at least one month per year in the next 30 years (Richey et al. 2015). The fundamental causes of water scarcity could be viewed in two different ways; scarcity in availability due to physical shortage, in this case natural circumstances are the main factors limiting access to usable water; e.g. very hot conditions that may imply low precipitation and decline in ground water-level, uneven distribution and seasonality of rainfall, relatively low stream flow in the rivers most of the time or the far distance of the points of use to the main water courses; it is therefore likely that water supply in such regions will often be overwhelmed by constantly exceeding water demands. The other form of water scarcity is the economic water scarcity which is predominant in developing countries; such form of water scarcity is due to the failure of institutions to ensure a regular supply or due to a lack of adequate infrastructure (United Nations 2015). The

E. Fosso-Kankeu (✉)
Department of Mining Engineering, College of Science Engineering and Technology, Florida Science Campus, University of South Africa, Johannesburg, South Africa
e-mail: fossoe@unisa.ac.za

T. T. I. Nkambule · B. B. Mamba
Institute for Nanotechnology and Water Sustainability (iNanoWS), College of Science Engineering and Technology (CSET), Florida Science Campus, University of South Africa, Johannesburg, South Africa

E. Fosso-Kankeu et al. (eds.), *Atmospheric Water Harvesting Development and Challenges*, Water Science and Technology Library 122,
https://doi.org/10.1007/978-3-031-21746-3_9

poor performance of wastewater treatment plants in most of the low-and-middle-income-countries contributes to high level of water wastage and pollution of rivers and groundwater with sewage, therefore significantly reducing the amount of usable water in the nearby communities.

Industrial activities have also been on the rise worldwide, contributing to the increase level and diversity of pollutants. The booming of the petroleum industry has resulted in the recurrent spillage of oil in the ocean and sea which results into an increase level of toluene in water affecting aquatic life and rendering water unsuitable for human use; it is estimated that about one million tons of oil makes its way into marine environments each year, coming mostly from consumers careless activities in factories, farms and cities. The effluents from the pharmaceutical industry are reported to contain high level of endocrine disruptors, antimicrobials and synthetic estrogens which contaminate freshwater near pharmaceutical plants and eventually end up in rivers, streams, lakes and oceans. A global-scale study focusing on environmental exposure to 61 active pharmaceutical ingredients (APIs) and carried out in 258 of the world's rivers across 104 countries, revealed that the most contaminated sites were the low-to-middle-income countries with poor wastewater and waste management infrastructure, while caffeine, carbamazepine and metformin were the most frequently detected APIs (Wilkinson et al. 2022). It is reported that every year around the world, about 60,000 tons of dyes are discharged in the environment as waste; while about 900 million tons of dyes contaminated wastewater from the textile industry are discharged into receiving water every year (Liu 2020). The mining industry is mostly responsible of the release of inorganic pollutants such as metals and sulphate into the environment; one of the main forms of pollution is through the formation of acid mine drainage with very low pH containing large amount of metals from the oxidation of exposed rock through mining activities (Fosso-Kankeu et al. 2020a).

A host of techniques have been used to address water pollution issues, these include biological, physical and chemical techniques which can be passive, active or hybrid. Some of the frequently used techniques are precipitation, reverse osmosis, ion exchange, activated sludge, anaerobic digestion, nanofiltration, advanced oxidation processes, adsorption and reducing and alkalinity producing system (RAPS) to name a few (Fosso-Kankeu et al. 2020b). These techniques may be effective for the treatment of some wastewater, but they have a certain number of limitations that often hamper their implementation at large scale; some of these challenges are high cost, production of toxic by-products, slow kinetics, energy intensive and therefore very costly (Fosso-Kankeu et al. 2015). Furthermore, the intensification of climate change is likely to render some of these techniques even less useful; It is expected that climate change will affect freshwater resources negatively, in term of both quality and quantity (Bagheri 2018). Therefore, the dependence on surface water at the abstraction points of drinking water plant, will render the plant less efficient as less water will be available while the pollutants will be further concentrated, increasing the complexity of water to be treated and putting more pressure on the available technologies. While predicting such situation in the near future, it is imperative to systematically look at alternative source of water that will be less challenged by the climate change.

Atmospheric water harvesting basically relies on atmospheric humidity as the principle behind the technology consists of capturing moisture from thin air, the captured moisture is then condensed into liquid water. Atmospheric water-harvesting technologies mostly make use of atmospheric water generators (AWGs) which can capture atmospheric moisture and condense into fresh water (Gido et al. 2016). Many countries in the world, and mostly those experiencing freshwater scarcity such as Arabic Gulf countries have been using this technology to supplement the shortage in their water resources. It is however important to mention that only little research on atmospheric water harvesters have been conducted to a comprehensive extent in order to facilitate the implementation of this technology in as many countries as possible. Currently the distribution of atmospheric water harvesting systems remain a huge challenge as only few systems are commercially operating. The market for atmospheric water generators can be segmented into three main classes namely industrial, commercial and residential. The main focus of this chapter is to provide up to date information regarding the market trend of products related to atmospheric water harvesting businesses in different regions around the world, breaking down the market size according to different sectors.

9.2 Atmospheric Water Harvesting Technologies

There are a number of atmospheric water harvesting technologies exploited around the world with the common approach of pulling water from the air. These technologies include among others, the water harvesting using condensation, sorption, fog catchers, and the sunlight.

The use of atmospheric water generators is mainly based on two techniques namely sorption technology and condensation technology. Active refrigeration technologies such as adsorption/absorption refrigeration, vapor compression cycle (VCC) and thermoelectric cooling (TEC) are often needed for the condensation technology. It is important to note that for condensation technology, electricity is needed to power the TEC and VCC. For the water vapor to be condensed, the cooling source temperature should be lower than the dew point temperature of the incoming air (Tu and Hwang 2020).

For the sorption technology, spongey adsorbents or absorbents are used to passively catch airborne water molecules from the atmosphere and turning them into liquid water without necessarily needing an external power source or moving parts. The type of sorption material establishes the difference between adsorbent technology and absorbent technology. A number of desiccants have been used in the sorption technology including calcium chloride, silica gel, activated carbon and lithium chloride (Liu et al. 2016). However, recently molecular organic framework has demonstrated attracting potential for water harvesting with the advantages of large water uptake and low temperature desorption. Logan et al. (2020) carried out a parametric study on nine hydrolytically stable MOFs with diverse structures for unravelling fundamental material properties. They therefore assessed the effects of relative humidity, powder bed thickness and temperature on the adsorption–desorption process. Their findings claimed a record performance for any reported MOF-based system, as they achieved $8.66 L_{H_2O}/kg_{MOF}/day$ production using Zr-MOF-808.

Fog harvesting using fog catchers/collectors is a technique viable in areas where fog is naturally generated in reasonable amount. The fog generally forms when the saturated water vapor condenses into small water droplets as the temperature decreases to a certain extent. The droplets gradually grow to critical size when they are captured on the net and then transported into the fog collectors which are mostly composed of network structure. An example of fog collector is the Raschel mesh which is based on textile weaving technology and has been widely used for fog harvesting due to many reasons; including the perfect anisotropic mechanical behavior, the simple fabrication and the low cost (Carrera-Villacres et al. 2020; Cheng et al. 2021; Yu et al. 2022).

Hydropanels is rather a sustainable technique that mostly rely on the sun for energy needed to absorb water vapor from the air. The air passing through a water-absorbing material is pulled through using fans; the water vapor is then condensed into liquid that is collected and treated accordingly to produce water with acceptable quality for drinking purpose. This technology is used in many countries around the world to supply clean drinking water to the population. A typical example is the installation of 15 rapid access clean water systems to Navajo households in the United States of America (https://www.bikkurim.org/these-water-harvesting-technologies-transform-the-air-into-drinking-water-at-the-tap/).

9.3 Types of Businesses Around Atmospheric Water Harvesting

Although the practice of collecting water from the fog dates several centuries ago, it is only until the twentieth century that the concept of experimental condensers becomes popular and some attempts to build them were materialised. The initial work by the Russian FI Zibold on the experimental stone condenser inspired several other research which eventually resulted in a significant finding on dew research by Monteith in 1957 which allowed a better understanding of the energy and heat balance mechanisms of dew formation and dew evaporation (Tu and Hwang 2020). The continuous work of researchers around the world has contributed to the realization of modern atmospheric water harvesting technologies such as atmospheric water generators which were initially mainly used in the arid and semi-arid areas, but are now extensively used around the world to complement the conventional water supply system. The development of atmospheric water harvesting technologies is also a catalyst for many businesses such as the commercialization of atmospheric water generators, desiccants, hydropanel, fog collectors and bottled drinking water.

9.3.1 Businesses Related to Atmospheric Water Generators

Atmospheric water generators which have the potential to draw moisture from atmospheric air, are considered as the main drivers of the atmospheric water harvesting technology. Such systems rely on latent heat and are therefore dependent on the temperature and relative humidity of the ambient air. With the climate change that is manifested by continuous increase of temperature, there has been steady fall in the availability of fresh water leading to an increase demand for atmospheric water generation. In this context AWG offers an economical alternative to bottled water as the cost of water produced using AWG is 50% less than that of the conventional bottled water. However, for most households and light commercial applications, the acquisition of AWG is still beyond reach due to the relatively high cost which also negatively affects the growth of AWG market. The end users of AWG include residential, commercial and industrial; in 2019 the largest share (accounting for about 76.3% of the global revenue) in the global market of atmospheric water generator was contributed by the industrial segment for which AWG is primarily employed in building and industries requiring 5000 gallons per day water output. Industries account for a considerable share (about 19%) of the global water consumption which is expected to further increase with the growing population; as climate change and pollution are putting more pressure on the limited inland freshwater, governments have put in place stringent regulations to limit the amount of water withdrawn from the rivers by industries hence the tendencies for these industries to consider sustainable and economical water solutions such as AWG to complement their traditional source. In 2019, the global market size for atmospheric water generator was estimated at USD 2.1 billion, while in North America where the US dominates the demand for AWG units, the market share was estimated to about USD 217.8 Million for the same period. There are currently two types of AWG products on the market, namely wet desiccant and cooling condensation. Because of its large-scale installations in various industries, commercial and residential buildings, cooling condensation is considered as the most widely used technique, estimated to account for about 98.9% share of the global revenue in 2019. The major components of cooling condensation AWG include the condenser, the compressor, the humidifier and the evaporator. The wet desiccation has a lower market share (around 17.8%) which is probably due to its complex mechanism as well as the ignorance of consumers with regard to wet desiccation AWG (https://www.grandviewresearch.com/industry-analysis/global-atmospheric-water-generator-market).

The global atmospheric water generator market is dominated by a wide range of large and small scale manufacturers from various regions that can be segregated into North America, Asia Pacific, Europe, Latin America, and Middle East and Africa (Table 9.1). While the Asia Pacific countries are already considered as major players

Table 9.1 Key companies operating in global atmospheric water generator market

Company	City	country	Region
Drinkable Air Technologies	Florida	USA	North America
Water Technologies International Inc	Florida	USA	
Atmospheric Water Generator LLC	Florida	USA	
Atlantis Solar	New York	USA	
Island Sky Corporation	Florida	USA	
Dew Point Manufacturing	British Columbia	Canada	
Eurosport Active World Corporation	Florida	USA	
Water-Gen Ltd	North Carolina	USA	
Clean Wave Products		USA	
Air2Water LLC	Los Angeles	USA	
EcoloBlue Inc	California	USA	
PlanetsWater Ltd	London	UK	Europe
GENAQ Technology SL	Lucena	Spain	
Eshara Water	Abu Dhabi	UAE	
Energy and Water Development Corp	Hamburg	Germany	
Saisons Trade & Industry Private Limited	Mumbai	India	Asia Pacific
Zhongling Xinquan (Fujian) Air Drinking Water Technology Co Ltd	Fujian Province	China	
WaterMaker India Pvt Ltd	Maharashtra	India	
Akvosphere	West Bengal	India	
Watergen	Petah Tikva	Israel	Middle East & Africa

on the market, it is predicted that the Middle East and Africa countries will eventually emerge as major players in this sector due to drastic water scarcity in some of these countries; however, for most of African countries the challenge remain their weak economic capacity as in many of these countries, the majority of the population will not be able to afford the required capital to invest in atmospheric water generator units.

9.3.2 *Desiccants Market*

Desiccants are used in adsorption-based technologies where they are saturated with water from the air then low-grade energy such as solar energy is used to desorb the adsorbed water which is then condensed into droplets collected in liquid form. A wide range of desiccants have been used for research or for practical application in the industry, among them are the silica gel, the polymers, calcium chloride, activated carbon fiber, zeolites, lithium chloride and recently metal organic framework. The performance of these materials could be assessed based on their adsorption and desorption capacities, but also their regeneration potential. Zeolite and silica gel have been reported to have weak adsorption capacities (https://absortech.com/knowledge/why-calcium-chloride/#:~:text=Calcium%20chloride%20desiccant%20absorbs%20more,like%20silica%20gel%20and%20clay). The adsorption capacity of calcium chloride desiccant which is made of hydrochloric acid and calcium carbonate is reported to be several times that of silica gel. Calcium chloride is available on the market in the form of bags and according to the Chinese based company Minghui, it has the following properties: non-toxic, fast adsorption kinetic, high adsorption capacity, non-environmental pollution, harmless to human body, odorless, non-contact corrosive and static dehumidification.

According to the Allied Market Research, although the global desiccant industry generated only $0.9 billion in 2020, it is expected that the desiccant market will grow to garner $1.5 billion globally by 2030. This prediction is essentially based on the following desiccant types, activated charcoal, clay, calcium chloride, silica gel, activated alumina and zeolite, considering end users such as packaging, electronics, pharmaceutical and food. The dominance of silica gel in this market is due to its preference by the pharmaceutical industry which is one of the key player in the desiccant market; it is estimated that silica gel segment accounted for the highest market share in 2020, with about two-fifth of the global desiccant market. According to current prediction, silica gel is expected to dominate the desiccant market until 2030. The global silica gel market is dominated by Asia–Pacific mostly because of its fast-growing packaging industry; while North America which occupies the second place will experience increase of the market size due to continuous growth of research activities in pharmaceuticals and biopharmaceuticals sectors.

The desiccants market in general is dominated by the packaging industry which drives the growth in the desiccants market. The major companies in the desiccants market include Zeotec Adsorbents Private Limited, Fuji Silysia Chemical, Dow Chemical, Desicca Chemicals, INEOS, Porocel, Qingdao Makll and Hengye Molecular Sieve. The desiccants market is segmented in about seven regions which include North America, Asia–Pacific, Western Europe, Eastern Europe, South America, Middle East and Africa. In 2021, Asia–Pacific was reported as the largest region in the desiccants market followed by North America. In these regions, a number of countries could be considered as the major players on the desiccants market, and these include the USA, China, India, Japan, Australia, Germany, UK, France, Indonesia, Russia, South Korea and Brazil (https://www.thebusinessresearchcompany.com/report/desiccants-global-market-report).

For the segmentation of the global desiccants market, a number of criteria including the type, the application and the process could be considered as shown in Table 9.2.

Table 9.2 Key segments of global desiccants market

Segmentation of desiccant market	
Type	Activated charcoal, clay, zeolite, silica gel, calcium chloride, activated alumina, others
Process	Physical adsorption, chemical adsorption
Application	Packing, pharmaceutical, food, electronics, air and gas drying, others

9.3.3 Hydropanels Market

Atmospheric water harvesting using solar hydropanels is a concept invented by researchers at the Arizona States University in the USA (Ferwati 2019; Tanavade et al. 2020). The hydropanels consist mainly of two separate parts, one that can adsorb moisture from the air while the other can generate heat. The system is sustainable as it can function off-grid relying only on the sun and the air to produce water safe for drinking purpose. The water that is produced after condensation will be mineralized with magnesium and calcium to improve the taste.

The market of hydropanels is mainly dominated by a company named Source Global which was created in 2015 by Cody Friesen who is the inventor of the technology. Source Global formally named Zero Mass Water has installed hydropanels in about 52 countries around the world in 450 separate projects. The company has gained from Bill Gates through collaboration with his companies such as Duke Energy, Breakthrough Energy Ventures and Blackrock, raising $150 million from investors (https://www.cnbc.com/2022/03/28/bill-gates-and-blackrock-backing-source-global-maker-of-hydropanels.html). The market segmentation of hydropanels is relatively broad and include commercial, residential, public, farming and remote applications. The driving force behind the market of hydropanels is mainly its sustainability and the potential to produce in almost any part of the world; however, the capital cost which currently is estimated at about $2000 per unit restrains the affordability for population in many countries. It is however expected that with the growing mass production, the unit price will eventually decrease, and the market size will consequently increase. So far, charitable initiatives have been responsible of the financial support of most of the projects on the installation of Source hydropanels in developing countries. Examples of such projects include the Musenga Vhadzimu Primary School in South Africa, USAID projects for refugees in Syria and Lebanon, and aboriginal communities in Australia, the Samburu Girls Foundation in Kenya. In the case of the Musenga Vhadzimu Primary School, the project was founded by New York philanthropist Malaak Compton-Rock which spent around $135,000 for the acquisition of 45 hydropanels donated to the primary school (https://www.globalwaterintel.com/global-water-intelligence-magazine/21/5/general/selling-water-at-150-m3-to-the-world-s-poorest-people-with-billionaire-backing) (Fig. 9.1).

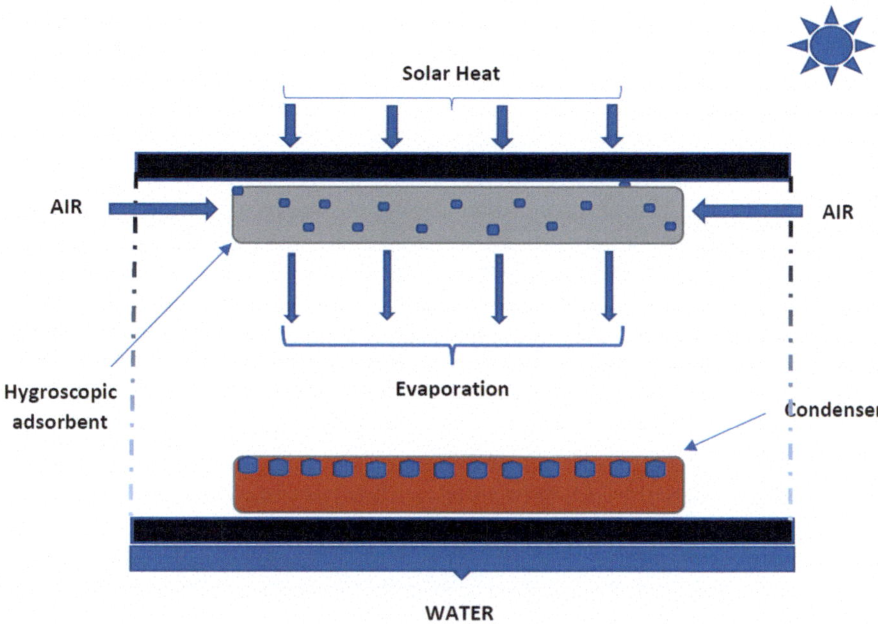

Fig. 9.1 Solar hydropanel system producing water from air and solar energy

9.3.4 Fog Collectors

Fog collection consists of inducing the fog-droplets by using large pieces of vertical mesh netting often known as fog collector or fog net, larger water droplets are formed as the minute fog droplets coalesce at the surface of the mesh (Fessehaye et al. 2014). The process has a myriad of advantages which include long service life, low economic cost and environmental friendliness which have contributed an increasing attention from the population over the world in the past few decades. The mesh used as collection derived from a textile weaving technology; one of the popular mesh named Raschel mesh is limited by its single function which significantly affects its fog harvesting efficiency (Rajaram et al. 2016). The cost of establishing fog collection systems depends mainly on the piping, water tanks, other equipment and supplies, labour availability and the material price. The cost of the mesh is influenced by the type of fabric, while to express the cost of fog collection system, people often consider the m^2 of the mesh installed for fog collection. For example, the two-dimensional Raschel mesh which is used for collection systems and considered as one of the most affordable systems, will cost between \$25 and \$50/m^2 of mesh (LeBoeuf and Jara 2014; Holmes et al. 2014; Fessehaye et al. 2017; Qadir et al. 2018).

Fog collection projects have been carried out in many areas around the world (Table 9.3), specifically area with limited access to fresh water. In the region of South and Central America, one of the interesting projects was in the Guatemalan village of Tojquia in 2006 where around 35 large fog collectors (LFC) were installed and produced an average 6300 L of water per day during the winter period (Henderson and Falk 2001; Schemenauer et al. 2007). In Europe, an example of fog collection project was in Croatia on the Mount Velebit which showed good potential as water source especially during the dry summer season (Mileta and Likso 2010). In the African region, the project was carried in two schools in the Soutpansberg Mountains of the Limpopo Province in South Africa where seven LFCs were installed to provide drinking water between 2001 and 2008 (Olivier and Heerden 2003). In the Asia region, a good example of fog collection implementation is the case of Nepal where the first installation dates back in 1997 as they used SFCs for evaluation purposes (MacQuarrie et al. 2001); the project was supported by NGO staff who were adequately trained to ensure effective running and maintenance of the system.

Compared to other water supply systems, the technology of fog collection is relatively cheap as is does not require energy and the operational cost is low. Fog collection systems have mostly been implemented in remote or rural areas of developing countries where the communities do not have the financial means necessary to install and maintain such systems; although the cost to install LFC may vary depending on the site access, it is estimated that the installation of 100 LFC units will cost about USD 40,000. A typical case study for a project carried out in Chungungo (Chile) where the cost of installing 60 LFC units was analysed, showed that about USD37000 was spent considering 6.2 km pipeline and 100 m^3 storage tankers. Since the communities in most instances cannot afford such costs, the majority of reported projects are thus sponsored by international organisations which have taken the initiative to train local staff for the maintenance of the systems (Fessehaye et al. 2014; Schemenauer et al. 1988).

9.3.5 *Bottled Atmospheric Drinking Water Market*

About 3 billion people are affected by water shortage in this planet. With the consistent growth of the population, the margin between water supply and demand is expected to increase even bigger in the coming years. Through the sustainable development goal 6 (SDG-6), member states of the UN have committed to ensure universal access to safely managed water which is fundamental for human health, socioeconomic development, wellbeing and gender equity. For many countries facing physical water scarcity such as the countries in the Middle East and North Africa, they have no option than to produce their own water. The consumption of fresh water is relatively high in certain countries; China consumed around 599 billion m^3 in 2018, Australia water consumption was 16, 132 GL in 2015–2016, while South Africa consumes about 16 billion m^3/year. It is important to mention that the agricultural sector and industries are the major consumers of water. However, the access to safe drinking water remains a challenge for most of developing countries with large populations and poor maintenance of public infrastructures, which contributes to the rise of bottled drinking water market. Between 2018 and 2019, the total amount of

Table 9.3 Fog collection systems installed around the world

Region	Location	LFC capacity	Period	References
South and Central America	Coquimbo	50 FC of 48 m^2 collection surface each	1980s	Schemenauer et al. (1988)
	Padre Hurtado/Chile	Water for chruch canctuary and gardening	1999–2004	Osses et al. (2000)
	Alto Patache/Chile	6 L/m^2/d	2002	Larrain et al. (2002)
	Mejia/Peru	Water for afforestation	1995–1999	Schemenauer and Cereceda (1993)
	Pachamama Grande/Ecuador	40 LFC of 12 L/m^2/d	1995–1997	Henderson and Falk (2001)
	Tojquia/Guatemala	35 LFCs for 6300 L/day	2006	Henderson and Falk (2001)
Africa	Soutpansbergs Mountains/South Africa	7 LFCs for 10 L/m^2/d	2001–2008	Olivier and Heerden (2003)
	Namib Desert/Namibia	14 SFCs	1998	Mtuleni et al. (1998)
	Nefasit & Arborobo/Eritrea	20 LFCs		www.fogquest.org
	Oman	30 L/m^2/d	1989–1990	Schemenauer and Cereceda (1992)
	Hajja/Yemen	25 LFCs for 4.5 L/m^2/d	2003–2004	Schemenauer et al. (2004)
	Bica de Cana/Madeira	8 L/m^2/d	2007	Prada et al. (2007)
	Tenerife Island	10 L/m^2/d	1998	Marzol and Valladares (1998)
	Sdesert Archipelago/Cape Verde	3–75 L/m^2/d	2004	Sabino (1998)
	NW Coast of Africa	7 L/m^2/d	2006	Marzol and Sanchez (2008), Marzol et al. (2010)
Europe	Mount Velebit/Croatia	4 L/m^2/d	2000	Mileta (2001), Mileta (2004), Mileta (2007)
	Iberian Peninsula/Spain	24 LCs for 7 L/m^2/d each	2003	Estrela et al. (2008)
	Mount Machos/Spain	3.3 L/m^2/d	2007	Estrela et al. (2009)

(continued)

Table 9.3 (continued)

Region	Location	LFC capacity	Period	References
Asia	Nepal	6 FCs of 20–30 m^2 mesh area	1997	MacQuarrie et al. (2001)

bottled water volume in the US increased from 13.8 billion gallons to 14.4 billion gallons according to the Beverage Marketing Corporation in the US. Market predictions estimated that by 2021, global bottled water market was going to reach around \$350 billion. The key players operating in the global drinking water market include Gerolsteiner Brunnen GmbH & Co. KG, Reignwood Investments UK Ltd., Bisleri International Pvt. Ltd., The Coca-Cola Company, Nestle S.A., Hangzhou Wahaha Group Co., PepsiCo, Inc., The Wonderful Company LLC, Icelandic Water Holdings hf (https://www.fortunebusinessinsights.com/drinking-water-market-106205). The global bottled water market is dominated by the Asia Pacific region, followed by Americas, Europe and the Middle East and Africa.

Technological innovation and advancement have resulted in the emergence of companies producing drinking water from the air. Atmospheric water generators offer an economical alternative to bottled water as the cost of water produced using AWG is 50% less than that of the conventional bottled water. Examples of such companies producing bottled drinking water from air include Atmospheric Water Solution from the US, Drupps in Sweden and Aquasky in South Africa. There are limited reports regarding bottled drinking water generated from the atmosphere, hence the market size could not be estimated in this review.

9.4 Conclusion

The gap between water demand and supply is ever growing while the impact of climate change is continuously affecting and reducing our limited water sources. As a host of countries around the world are already overwhelmed by the challenges of economic water scarcity as they struggle to maintain their water infrastructure; it is expected that things will even worsen as the quantity of fresh water is considerably reduced in the near future. In some parts of the world there have been proactive research in anticipation of foreseen water crisis coupled with climate change. Atmospheric water harvesting offers the advantage not to rely on surface water sources as water is extracted from the air. This approach could therefore be the best option to produce water under circumstances of low precipitation or extreme physical water scarcity. Hence the recent development in technologies related to atmospheric water harvesting, including atmospheric water generators, fog collection and hydropanels. However, the market trends clearly show that a huge share of the market belongs to developed countries in the Asia Pacific, North America and Europe where the end users of AWG include residential, commercial and industrial; however, in such countries access to safe fresh water by the population is not a major concern. It is expected that the market of AWH technologies will significantly grow in the near future as the world experience a significant population growth coupled with climate change. The low-and-middle-income-countries have been benefiting from the AWH technologies through charitable initiatives which have been responsible of the financial support of most of the projects on the installation of fog collection systems and hydropanels. The development of AWGs has also contributed to significantly lower the cost of the bottled drinking water, hence the recording of considerable growth in this market. The continuous development of AWH technologies is expected to contribute in the lowering of the installation cost rendering these technologies affordable for developing countries which are the most affected by fresh water scarcity.

References

Bagheri F (2018) Performance investigation of atmospheric water harvesting systems. Water Resour Ind 20:23–28

Carrera-Villacres D, Carrera-Villacres JL, Braun T, Zhao Z, Gomez J, Quinteros-Carabali J (2020) Fog harvesting and IoT based environment monitoring system at the Ilalo volcano in Equador. Int J Adv Sci Eng Inform Technol 10(1):407–412

Cheng Y, Zhang S, Liu S et al (2021) Fog catcher brushes with environmental friendly slippery alumina micro-needle structured surface for efficient fog-harvesting. J Clean Prod 315:127862

Estrela MJ, Valiente JA, Corell D, Millan M (2008) Fog collection in the western Mediterranean basin (Valencia region, Spain). Atmos Res 87:324–337

Estrela MJ, Valiente JA, Corell D, Fuentes D, Valdecantos A (2009) Prospective use of collected fog water in the restoration of degraded burned areas under dry Mediterranean conditions. Agric for Meteorol 149:1896–1906

Ferwati FM (2019) Water harvesting cube. SN Appl Sci 1:779. https://doi.org/10.1007/s42452-019-0730-y

Fessehaye M, Abdul-Wahab S, Savage MJ, Kohler T, Gherezghiher T (2014) Fog-water collection for community use. Renew Sustain Energy Rev 29:52–62

Fessehaye M, Abdul-Wahab SA, Savage MJ, Kohler T, Ghereghiher T, Hurni H (2017) Assessment of fog-water collection on the eastern escarpment of Eritrea. Water Int 42:1022–1036

Fosso-Kankeu E, Waanders FB, Steyn FW (2015) The preparation and characterization of clay-biochar composites for the removal of metal pollutants. In: Muzenda E, Yingthawornsuk T (eds) 7th international conference on latest trends in engineering and technology (ICLTET' 2015), Nov 26–27 2015, Irene, Pretoria (South Africa). ISBN: 978-93-84422-58-5

Fosso-Kankeu E, Wolkersdorfer C, Burgess J (2020a) Recovery of by products from acid mine drainage treatment. Wiley Scrivener. ISBN: 978-1-119-62018-1

Fosso-Kankeu E, Pandey S, Ray SS (2020b) Photocatalysts in advanced oxidation processes for wastewater treatment. Wiley Scrivener. ISBN: 978-1-119-63139-2

Gido B, Friedler E, Broday DM (2016) Assessment of atmospheric moisture harvesting by direct cooling. Atmos Res 182:156–162

Henderson B, Falk D (2001) Fog water collection in Ecuador: an appropriate technology for the rural poor. In: Proceedings of the 2nd international conference on fog and fog collection. 15–20 July 2001, Vancouver, Canada, pp 281–284

Holmes R, Rivera JD, de la Jara E (2014) Large fog collectors: new strategies for collection efficiency and structural response to wind pressure. Atmos Res 151:236–249

https://absortech.com/knowledge/why-calcium-chloride/#:~:text=Calcium%20chloride%20desiccant%20absorbs%20more,like%20silica%20gel%20and%20clay

https://www.bikkurim.org/these-water-harvesting-technologies-transform-the-air-into-drinking-water-at-the-tap/

https://www.cnbc.com/2022/03/28/bill-gates-and-blackrock-backing-source-global-maker-of-hydropanels.html

https://www.fortunebusinessinsights.com/drinking-water-market-106205

https://www.globalwaterintel.com/global-water-intelligence-magazine/21/5/general/selling-water-at-150-m3-to-the-world-s-poorest-people-with-billionaire-backing

https://www.grandviewresearch.com/industry-analysis/global-atmospheric-water-generator-market

https://www.thebusinessresearchcompany.com/report/desiccants-global-market-report

Larrain H, Velasquez F, Espejo R, Pinto R, Cereceda P, Osses P, Schemenauer RS (2002) Fog measurements at the site "Falda Verde" north of Chanaral compared with other for stations of Chile. Atmos Res 64:273–284

LeBoeuf R, de la Jara E (2014) Quantitative goals for large-scale fog collection projects as a sustainable freshwater resource in northern Chile. Water Int 39:431–450

Liu Q (2020) Pollution and treatment of dye waste-water. Earth Environ Sci 514:052001

Liu JY, Wang JY, Wang LW, Qi Y (2016) Performance test of sorption air-to-water device. CIE J 67(S2):46–50

Logan MW, Langevin S, Xia Z (2020) Reversible atmospheric water harvesting using metal-organic frameworks. Sci Rep 10:1492

MacQuarrie KIA, Pokhrel A, Shrestha Y, Osses P, Schemenauer RS, Vitez F, Kowalchuk K, Taylor R (2001) Results from a high elevation fog water supply project in Nepal. In: Proceedings of the 2nd international conference on fog and fog collection. Vancouver, Canada, pp 227–229, 15–20 July 2001

Marzol MV, Sanchez J (2008) Fog water harvesting in Ifni, Morocco. An assessment of potential and demand. Erde 139:97–126

Marzol MV, Sanchez J, Yanes A, Derhem A, Bargach J (2010) Meteorological patterns and for water in Morocco and the Canary Islands. In: Proceedings of the 5th international conference on fog. Fog Collection and Dew, Munster, Germany, pp 56–59, 25–30 July 2010

Marzol MV, Valladares P (1998) Evaluation of fog water collection in Anaga (Tenerife, Canary Islands). In: Proceedings of the first international conference on fog and fog collection. Vancouver, Canada, pp 449–452, 19–24 July 1998

Mileta M (2001) Fog water collection at the Mountain Velebit near the Adriatic Sea. In: Proceedings of the 2nd international conference on fog and fog collection. Vancouver, Canada, pp 265–268, 15–20 July 2001

Mileta M (2004) Results from fog water water collection on Mt Velebit in Croatia. In: Proceedings of the third international conference on fog. Fog Collection and Dew, Cape Town, South Africa, 11–15 Oct 2004

Mileta M (2007) Seven years of fog measurements with SFC in Croatia. In: Proceedings of the 4th international conference on fog. Fog Collection and Dew, La Serena, Chile, pp 29–32, 22–27 July 2007

Mileta M, Likso T (2010) Fog water collection with SFC on the mountain Velebit (Croatia) during the period 2000–2009. In: 5th international conference on fog. Fog Collection and Dew Munster, Germany, 25–30 July 2010

Mtuleni V, Henschel JR, Seely MK (1998) Evaluation of fog-harvesting potential in Namibia. In: Proceedings of the first international conference on fog and fog collection. Vancouver, Canada, 179–182, 19–24 July 1998

Olivier J, van Heerden J (2003) Implementation of an operational prototype fog water collection system. Water Research Commission Report No. 902/1/02, 91pp

Osses P, Schemenauer S, Cereceda P, Larrain H, Correa C (2000) Los atrapanieblas del Santuario Padre Hurtado y sus proyecciones en el combate a la desertification. Revista Geografia Norte Grande 27:61–67

Prada S, Oliveira da Silva M, Figueira C, Meneses M, Pontes A (2007) Fog water collection in Madeira Island (Portugal). In: Proceedings of the 4th international conference on fog. Fog Collection and Dew, La Serena, Chile, pp 277–283, 22–27 July

Qadir M, Jimenez GC, Farnum RL, Dodson LL, Smakhtin V (2018) Fog water collection: challenges beyond technology. Water 10:372

Rajaram M, Heng X, Oza M, Luo C (2016) Enhancement of fog-collection efficiency of a Raschel mesh using surface coatings and local geometric changes. Colloids Surf, A 508:218–229

Richey AS, Thomas BF, Lo M-H, Reager JT, Famiglietti JS, Voss K, Swenson S, Rodell M (2015) Quantifying renewable groundwater stress with GRACE. Water Resour Res 51(7):5217–5238

Sabino A (1998) Fog water collection in Cape Verde Islands: an alternative source of water for agriculture and domestic use. In: Proceedings of the first international conference on fog and fog collection. Vancouver, Canada, pp 445–448, 19–24 July 1998

Schemenauer RS, Cereceda P (1992) Monsoon cloud water chemistry on the Arabian Peninsula. Atmos Environ 26A:1583–1587

Schemenauer RS, Cereceda P (1993) Meteorological conditions at a coastal fog collection site in Peru. Atmosfera 6:175–188

Schemenauer RS, Cereceda P, Fuenzalida H (1988) A neglected water resource: the camanchaca of South America. Bull Am Meteor Soc 69:138–147

Schemenauer RS, Osses P, Leibbrand M (2004) Fog collection evaluation and operational projects in the Hajia Governorate, Yemen. In: Proceedings of the third international conference on fog. Fog Collection and Dew, Cape Town, South Africa, 11–15 Oct 2004

Schemenauer RS, Rosato M, Carter V (2007) Fog collection projects in Tojquia and La Ventosa, Guatemala. In: Proceedings of the 4th international conference on fog. Fog Collection and Dew, La Serena, Chile, pp 383–386, 22–27 July 2007

Tanavade S, Manic S, Al-Khazraji A, Charkaoui A (2020) Water from sun: an energy conservation initiative for smart cities. In: Proceedings of the 3rd IET international smart cities symposium, 3rd-SCS-2020, 21–23 Sept 2020, Bahrain

Tu R, Hwang Y (2020) Reviews of atmospheric water harvesting technologies. Energy 201:117630

United Nations (2015) The United Nations world water development report 2015. Water for a sustainable world. The United Nations Educational, Scientific and Cultural Organization, 7, place de Fontenoy, 75352 Paris 07 SP, France

Wilkinson JL, Boxall ABA, Kolpin DW et al (2022) Pharmaceutical pollution of the world's rivers. Proc Natl Acad Sci USA 119(8):2113947119

www.fogquest.org

Yu Z, Zhu T, Zhang J, Ge M, Fu S, Lai Y (2022) Fog harvesting devices inspired from single to multiple creatures: current progress and future perspective. Adv Func Mater. https://doi.org/10.1002/adfm.202200359

Chapter 10
Awareness of Atmospheric Water Harvesting Technology in a Community: Case Study of Pretoria North in South Africa

Palesa Mkabane, Frans Boudewijn Waanders, Elvis Fosso-Kankeu, Ali Al Alili, and Hemant Mittal

10.1 Introduction

Access to clean water is a growing challenge globally, especially in areas where natural fresh water is not available, and communities rely on processes like desalination for clean water. In 2015, South Africa had the lowest amount of rainfall since records began in 1904. While it is predicted future climate conditions will be worse, alternative crops, water sources and methods of electricity production will need to be considered; failing to do this will result in increased food, energy, and water insecurity (Piesse 2016). Atmospheric water harvesting (AWH) is proposed in this study as an alternative water source in South Africa; as both the source of choice during natural disasters and with prevailing water pollution effects due to industrialisation and urbanisation.

Atmospheric water harvesting (AWH) continues to gain momentum especially in the recent years as the earth undergoes challenges with water limitations due to global warming effects and water pollution. Though it is known amongst people that plenty of water exists in the air they breathe; the fundamental limits, successes and

P. Mkabane (✉) · F. B. Waanders
Water Pollution Monitoring and Remediation Initiatives, School of Chemical and Minerals Engineering, Center of Excellence in Carbon-Based Fuels, North-West University, Private Bag X1290, Potchefstroom 2520, South Africa
e-mail: mkabanept@gmail.com

E. Fosso-Kankeu (✉)
Department of Mining Engineering, College of Science Engineering and Technology, Florida Science Campus, University of South Africa, Johannesburg, South Africa
e-mail: elvisfosso.ef@gmail.com

A. Al Alili · H. Mittal
Dubai Electricity and Water Authority (DEWA), DEWA R&D Center, Dhabi, United Arab Emirates

E. Fosso-Kankeu et al. (eds.), *Atmospheric Water Harvesting Development and Challenges*, Water Science and Technology Library 122,
https://doi.org/10.1007/978-3-031-21746-3_10

challenges of atmospheric water harvesting has to be thoroughly understood. When factual knowledge is available, then perspectives on how to search for a cost-effective way to produce atmospheric water can be achieved (Tu et al. 2018). Research on the topic has been diversified from harvesting air up to fog to produce water, which the two media are interlinked and are both referred to as AWH.

Fog is visible cloud water droplets suspended in the air at or near the earth's surface; whilst AWH is the collection of water vapour from air that has been cooled and condensed below its dew point temperature (Jarimi et al. 2020). Atmospheric water harvesting technologies can be achieved in two principles, namely condensation/vapour compression also known as active and absorption–desorption cycle known as passive; both technologies are solely based on the geographical location's climate. The location's climate is affected by its altitude, latitude, terrain, water sources and currents, consequently precipitation; hence the technology of AWH is based on temperature and humidity combination process whereby humidity is the quantity of water vapor present in air. Vapour compression is a process that employs a refrigerant which is circulated by a compressor through a condenser and an evaporator coil which cools the air surrounding it, lowering the air's dew point (Anbarasu and Pavithra 2011; Mkabane et al. 2020). Secondly, adsorption–desorption cycle is a process that allows natural absorption of water vapour in low temperatures i.e. evenings and during the day when the temperatures are warmer much higher than the water desorbs (removed or condensed) from an adsorbent (Kim et al. 2017).

The unique element about AWH is that the source of air is available everywhere whilst other methods of water harvesting require some availability of water for their success. Literature explains that whilst air is abundant, it requires an efficient process to capture and deliver water from air especially at low humidity levels i.e. 20%. An experiment of three residential-size AWH units of a nominal power of 1500 W were tested in laboratory environment by Bagheri (2018) under different simulation of climatic conditions namely: warm and humid, mild and humid, cold and humid, warm and dry, mild and dry, cold and dry, and mild. The experimental results showed that water harvesting yield increases by simultaneous increase of water content (ω) or dew point temperature (Tdewpoint) and a decrease in temperature. The average water harvesting rate varied in a range of 0.05 L/h for cold and humid conditions, to 0.65 L/h for warm and humid climates. The average energy consumption changed from 1.02 kWh/L for warm and humid to 6.23 kWh/L for cold and humid climates. Bagheri (2018) concluded his experiment with results that has proven that an ideal AWH condition was the one capable of cooling the air stream down to a point close to but above the freezing temperature (~ 1 °C) to achieve the highest water harvesting rate at all climates. The climatic comparisons showed an increase in the gaps between real and an ideal water harvesting rate through changing from low humidity water content condition to high humidity conditions.

The experiment further denoted a higher potential of performance improvement in higher humid climate conditions for optimal AWH. The experimental outcomes also indicated a failure of AWH systems to operate in cold-and-dry climate as well as a significant increase in energy cost of AWH in hot and dry conditions. The basis of Tdewpoint > 2 °C was developed and established from this experiment as a minimum climatic requirement for using AWH systems, any condition lower would not be worth the energy consumption as well as an amount of water required.

It is therefore evident that the performance of the AWH technologies is area specific, hence the type of technology and the associated cost, may considerably vary from one region to another and therefore impact on the acceptability of such a technology by the community. However, it is also important to mention that for some communities, the concept of harvesting water from the atmosphere or the modern way of doing it is relatively new and could therefore be regarded with a certain degree of scepticism, hence the need to assess through a survey, a community awareness and willingness to adopt such a technology for their daily water supply either during normal time or under the time of crisis. This was the aim of the proposed study that was carried out in both residential and industrial areas near Pretoria in South Africa.

10.2 Methodology

10.2.1 Study Area

This paper focuses on a case study investigating the acceptability of atmospheric water harvesting in a community situated on a global positioning system location 25.5864° S, 27.9876° E (Fig. 10.1), Pretoria North, Gauteng Province in South Africa.

10.2.2 Design

Surveys were conducted from local community members and surrounding towns for a period of a month; within 5 km radius from the case study site. The aim of the research survey was to assess and create awareness towards atmospheric water harvesting (AWH), considering that people and industries can opt for this type of water as an alternative from the usual potable water supply especially during infrastructure downtime, drought and other unforeseen circumstances. The survey shall further evaluate if people really care about the quality of water which consequently impacts on their health as well as the tariff costs associated to it.

Water in South Africa is calculated on a sliding scale, the more one consumes water, the higher the cost to pay. It was on this regard that the issue of quality versus tariff cost was highlighted through the engagement whilst noting usage that has to be monitored too. There are different water usage brackets whereby the billing starts at 0–9 kilolitre (kL) and it costs R16 per kL (Joburg 2020). Every once a month, on a specific date, a local municipal delegate takes readings from the water meter attached to a household or a business unit.

The participants contributed voluntarily to the research after being told that the research is about drinking water which is vital for human lives and socio-economic

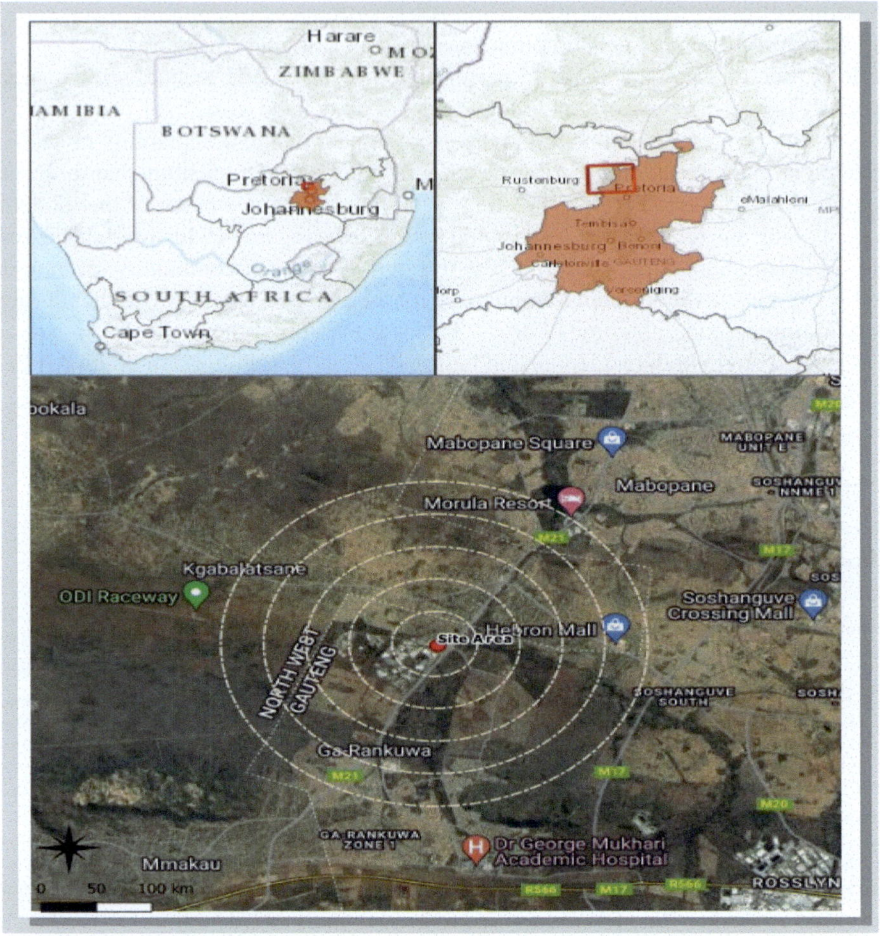

Fig. 10.1 Study area in Pretoria North

growth. Questionnaires and consent forms were provided to each person that agreed to participate in the research. No names, no race or identity of participants were captured; however, a consent form was attached with a count of a participant i.e. "participant 1" to track people's involvement for validity of randomization. Randomization was assured with reference to the principle of randomization by Ronald Fisher (DasGupta 2011) whereby a participant was engaged twice with the same questionnaire; within the same day, by two different people. The bias within participants depicts an error arising from the survey through characteristics such as sexuality, age, social class and beliefs of those who voluntarily participate in survey compared with those who do not. It has been observed from research that it is difficult to achieve outcome rates in excess of 80% although higher outcome rates has sometimes been observed from developing countries (Fenton et al. 2001).

The first set of interviews was conducted by the researcher whilst the second one was done by the person appointed by the researcher. Fieldworkers has to be highly vigilant during engagements with participants and have to be equipped with all necessary knowledge of how to interact with participants to ensure they earn their cooperation. The researcher chose the Masters student who was studying a similar topic as the fieldworker to conduct the survey with her, the advantage was that the student already understood the importance of information and ethics involved in conducting such a survey. The survey was conducted during the day between 8h00 until 16h00 hours, only during the week being any day between Monday and Friday. The places where engagements took place were at the taxi ranks, bus stops where these two places were suitable as people would be asked questions mostly when they are awaiting their transport in queues. Other places were a community hall on the day where government grants are being paid, on the industrial sites during lunch times where people would be in and out of their work places to get food, also at the mall entrances and a few times near the church.

Ronald Fisher method explains how the researcher can get quality statistical outcomes from their random surveys without being biased through the distribution surveys from participants' feedback; any difference spotted from the surveys can be explained and seen by the researcher. Only age shall be recorded for opinion validity purposes as any person above the age of 18 is considered an adult in South Africa (Government 2005). For this particular survey; opinions and validation matters as to what the people understood about drinking water challenges in South Africa and about atmospheric water (AW), the difference between AW and the usual potable water; if they will opt for AW as an alternative source to the normal potable water they are used to or not or use both these sources. The criteria of age were followed within the questionnaire which the data analysis distinguished that opinions were given from an adult thought with the following criteria:

- comparisons: conventional supply of drinking water versus atmospheric water supply.
- comparisons: water quality versus water costs.

The engaged community consists of one side being industrial area and the other being residential which was where most water users were. The main objective of the survey was to assess the awareness of AW technology to all water users and if the technology was worth investing on either by businesses or residences.

There are two types of selection of data methods namely qualitative and quantitative methods. Qualitative method is usually in sentences or words and is applicable when there is a gap and or challenge whilst there are no obvious available solutions to the challenge or the gap (Paradis et al. 2016), whilst the quantitative is either mathematical or numerical and applicable when concepts and theories has to be tested; the method applied for the community engagement survey of this project is the latter one.

10.2.3 Ethical Considerations

The researcher had to present the community engagement plan to the university for approval from the university's ethics committee. The committee had to assess the following amongst other information:

- type of the survey if it poses risks to humans or environment,
- the benefits of the survey,
- the criteria for participation recruitment,
- privacy and confidentiality of both data and participants,
- facilities or places where the survey took place,
- if there are incentives or not to the participants.

10.2.4 Execution Plan

The researcher circulated the prepared consent form to the ethics committee and research department for approval, the consent form highlighted the following information of a participant amongst other:

- the participation is voluntarily, the participant can withdraw at any given time without any prejudice,
- the participation is solely for research purposes,
- there is no remuneration attached to participation,
- no personal identification, sex, race except for age to distinguish between an adult and a child for survey information validity,
- the survey will be done twice for validity purposes.

The researcher circulated the prepared quantitative questionnaire to the ethics committee and research department for approval. The questionnaire had to be answered in a scale of **1 up to 5** whereby **1 = strongly disagree, 2 = disagree, 3 = neither agree nor disagree, 4 = agree and 5 = strongly** agree.

All participants had to first agree to the consent form, then mention their age to proceed to answer the questionnaire. The questionnaire had the following summary of questions to address the aim:

1. Do all South Africans have access to clean drinking water?
2. Is the drinking water supply infrastructure adequate in South Africa to ensure that drinking water reaches all intended places safely?
3. Is the drinking water supply infrastructure always available and reliable in South Africa?
4. Does the South African climate always allow enough rain during rainy season?
5. Have severe droughts been experienced in South Africa?
6. Are most water sources contaminated and polluted in South Africa?
7. Can atmosphere, fog and rain be harvested as drinking water source to assist water challenges gaps?

8. Are drinking water tariffs at a lower rate in South Africa compared to other countries?
9. Is drinking water quality more important than its cost?
10. Should the cost of drinking water be relative to the quality; can high quality water be costed higher while poor quality water should be cheaper?

Three meetings about community engagement took place between the researcher and her field worker to ensure that they do "dry run" for the preparations ahead of the survey. The main agenda points were the friendly approach to the targeted participant, she/he should agree first to be interviewed, confirm if the participant is an adult, take the participant through the consent form, then followed by the questionnaire and the message that another similar interview would follow in a few minutes. Literature mentions that when engaging with human beings its crucial to be friendly, humble, be respectful whilst maximising the benefit of time given for the engagement (Kabir 2016). The researcher has gone through ethics training with the university and completed an international academic ethics training which has equipped her with tools, concepts and methods to conduct this survey efficiently.

The researcher prompted the appointed field worker to conduct the survey on her as a mind preparation for the real community engagements. The researcher has also covered extensive literature on data collection during this time as a preparation of any unforeseen circumstance that could arise during the survey. The researcher was a few meters from the appointed field worker so that she could keep track of the person that has been interviewed, whilst the researcher informed the participant through the consent form that she/he will be interviewed by another person shortly when they are done with the particular interview.

10.3 Results

10.3.1 Participation Per Age Group

Table 10.1 shows that the major (40%) population group which participated during the survey engagement was the age group between 30 and 39 years old. This particular group was found mostly on the industrial site as well as the malls which suggests that the most influential group is from a working class. The group has buying power regarding change of water supply or even considering AWH as an alternative to what they are currently accessing. The survey team experienced challenges such as people not wanting to answer the questions mostly in the mornings during rush hours or time of their personal errands, the best time seemed to be during the day when people are relaxed and enjoying their lunch breaks too. What was also interesting was a little contribution of an elderly of 80–89 which really did not care or even bothered much however they were interested to know about the purpose and the reason for the study and if the changes would really come to reality in their lifetime. Younger people indicated as the initial group aged between 24 and 29 were highly cooperating

Table 10.1 Overall survey population participation

Age group	Survey contribution %
24–29	20
30–39	40
40–49	20
50–59	7
60–69	3
80–89	3

at any time of the day for the simple reason that some were students while others were unemployed, therefore having enough time on the side; it is also important to mention that is also a concerned group as they understand how such study could have a significant impact in their life for many years to come.

10.3.2 General Survey Outcome Regarding the Perception of the Different Age Groups

Table 10.2 depicts the results calculated from questionnaires as per the research questions that were formulated as mentioned in Sect. 2.2. The most favourable question was Question 9 where the majority (87%) of the participants feels that water quality is of an utmost importance than the tariff costs paid for it, while the least favourable was Question 1 whereby the majority (74%) of interviewed participants do not believe that "all South Africans have access to clean drinking water". Question 7 was also positively supported by most (80%) of the engaged people as they believed that atmospheric water, fog and or rain can be harvested for drinking water purposes. About 39% of interviewed participants believe that most water sources are polluted and contaminated in South Africa. Question 5 and 10 have been responded equally regarding water tariff costs being related to water quality supplied to consumers as well as the population's understanding about the droughts that have happened in South Africa.

10.3.3 Perception of Age Groups Regarding the Consideration of AWH

Figure 10.2 highlights the perception of age groups regarding an alternative water source (AWH) solution from the current source supplied by the government. Although the number of participants varies per age group, it is important to notice that in general, at least 66.6% of participant per group were in favour of an alternative source of water, in this case AWH. The negative answers were recorded from two

Table 10.2 Overall survey outcome results

Question	Scoring	1	2	3	4	5
		Strongly disagree (%)	Disagree (%)	Neither agree nor disagree (%)	Agree (%)	Strongly agree (%)
1		42	32	3	13	10
2		32	51	11	3	3
3		33	44	6	7	10
4		33	23	7	23	14
5		8	11	12	42	27
6		3	13	7	38	39
7		6	7	7	22	58
8		21	27	20	12	20
9		3	3	7	27	60
10		17	10	16	30	27

age groups namely 30–39 and 40–49; this is understandable as the participants from these age groups have just started families and are afraid of any instability as the AWH represents for them an unknown source they still have to discover and most of them tend to associate it with high cost. According to the literature, any number lower than 50% is less than the majority, in this case, one could say that there is acceptance of AWH as alternative source of water.

It is further explained in research that weak surveys could mean different reasons and cannot be concluded at first as failed survey or "non acceptability" as with this particular case. In-depth and inflexibility have been highlighted as the main

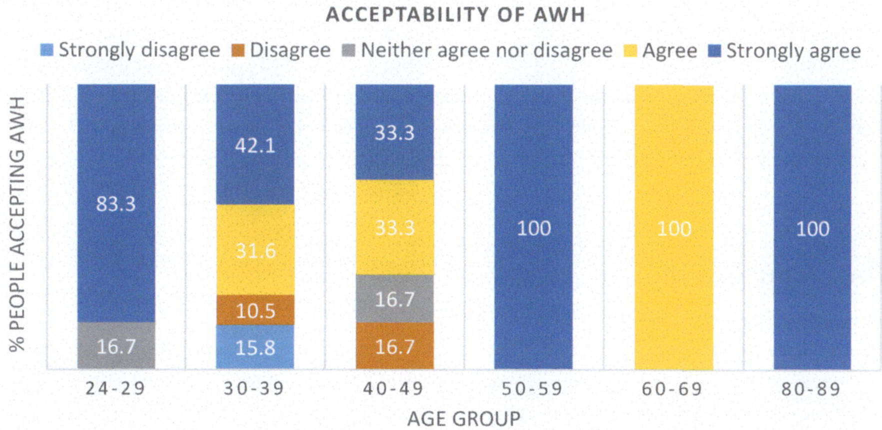

Fig. 10.2 Percentage of acceptance per age group interviewed (Q7 reflection)

contributor to lower results of the outcome (DeCarlo 2018). In-depth means that at a time a question was asked to the participant, he or she understood it in his or her own way at times with lack of knowledge on the subject and the researcher or filed worker then realises that the question could have been phrased differently as they realise, they did not achieve what they had anticipated. An issue of inflexibility could be addressed better on qualitative surveys as questions are narrated better in words and sentences in that way participants are able to express themselves better whilst the researcher can get a bigger picture of his or her survey data collection.

10.3.4 Perception of Age Groups Regarding the Payment for Quality Water

Figure 10.3 summarises the opinions of participants regarding their willingness to pay the required cost to get access to water of high quality. It is important to notice here that the views of respondents were quite diverse as a balance between positive and negative perception was recorded across the age groups. Apart from the two age groups 50–59 and 60–69 where 100% positive responses were recorded, one could observe that the majority of the participants in the age groups 24–29, 30–39, 40–49 and 80–89 did not agree to pay more money if they were supplied with water of higher quality. Our explanation for this trend is as follows, in the first age group (24–29) most of the people are unemployed, while in the age groups 30–39 and 40–49, the participants have just started a family or are experiencing the challenges of raising a family and therefore are less likely to consider additional cost in their monthly expenses; while people in the age group 80–89, rely on governmental grants which is very little for survival and automatically will not afford to pay more for any service.

Of all the questions within the survey, the question with the lowest support was number 2 with 3% of the participants believing that the South African drinking water infrastructure is adequate while 14% believe there is enough rain within the country. About 7% of the questionnaires were deemed not suitable for consideration in the study as the responses recorded by field worker from the same participant did not match those collected earlier by the researcher during the preliminary interview.

10.4 Conclusion

The survey engagement provided the reality of people's understanding concerning the current drinking water supply network, it's limitations mostly related to the supply infrastructure that are poorly maintained. Table 10.2 indicates that the majority of people according to the survey do not believe that the available drinking water infrastructure is adequate and that it is always available and reliable. The feedback provides

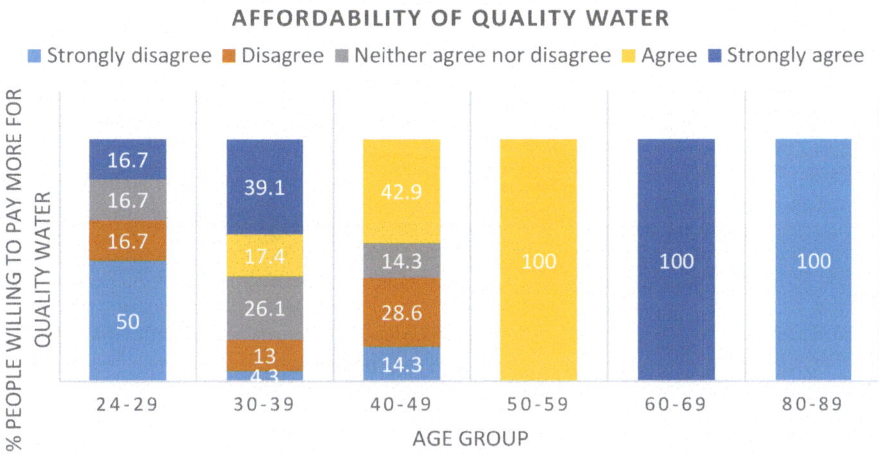

Fig. 10.3 Percentage of people willing to pay more for quality water (Q10 reflection)

a good opportunity for the introduction and expansion of AWH to water consumers to continually have access to clean drinking water whilst the government as the main supply stakeholder implements plans to regain trust and reinforce infrastructural integrity again.

The highest outcome of supported engagement was on the issue of water quality which was found to be highly crucial more than the cost of paying for the water itself as per question 9. It is evident that people are concerned about their livelihoods and the impact of drinking water on their lives; currently the selling price of water on average is R16 per kL (Joburg C.o 2020) whilst AW selling price is about twice the amount as per the costs from the site study business. The AW costs could be reduced by using cheaper energy sources such as the solar as most of the cost of production were impacted by the energy costs. The high-quality factor provided by AW could motivate more support from consumers even when selling price is relatively high compared to that of water supplied by municipalities.

It can be concluded from the overall survey results that people would be open to an alternative source of water judging by the responses of participants to Question 7. Only few did not consider the use of alternative source of water. The main challenge is however the affordability of quality water at relatively high cost; the views of respondents with regard to this aspect are mostly negative as they do not want to pay more for water of better quality. According to Stats SA, Statistics Department of South Africa, the highest employment rate age group is between 45 and 54, followed by 30–45, however, family and/or social responsibilities in these age groups can contribute to hamper the commitment of these participants to engage anything that will cost them more.

References

Anbarasu T, Pavithra S (2011) Vapour compression refrigeration system generating fresh water from humidity in the air. In: International conference on sustainable energy and intelligent systems (SEISCON, 2011)

Bagheri F (2018) Performance investigation of atmospheric water harvesting. Water Resour Ind 20:23–28. https://doi.org/10.1016/j.wri.2018.08.001

DasGupta A (2011) Selected works of Debabrata Basu, selected works in probability and statistics. Springer, pp 428. https://doi.org/10.1007/978-1-4419-5825-9_28

DeCarlo M (2018) Scientific inquiry in social work. https://creativecommons.org/licenses/by-nc-sa/4.0/

Fenton K, Johnson A, McManus S, Erens B (2001) Measuring sexual behaviour: methodological challenges in survey research. Sexual Transm Inf 77(2):84–92

Government R.o (2005) Children's Act 38. Government Gazette, 492

Jarimi H, Powell R, Riffat S (2020) Review of sustainable methods for atmospheric water harvesting. Int J Low-Carbon Technol 15:253–276. https://doi.org/10.1093/ijlct/ctz072

Joburg C.o (2020) Impact of water tariff increase on Joburg residences. Water Impact, Joburg Water

Kabir SMS (2016) Methods of data collection. In: Basic guidelines for research: an introductory approach for all disciplines. Book Zone Publication, Chittagong-4203, Bangladesh. Retrieved from https://www.researchgate.net/publication/325846997

Kim H, Yang S, Rao S, Narayanan S, Kapustin E, Furukawa H, Umans A, Yaghi O, Wang E (2017) Water harvesting from air with metal-organic frameworks powered by natural sunlight. Science 356:430–434

Mkabane P, Fosso-Kankeu E, Al Alili AR, Waanders F, Mittal H (2020) Water production yield of an atmospheric water generator during summer season in South Africa. In: Fosso-Kankeu E, Waanders F, Bulsara HP (eds) 18th Johannesburg international conference on science, engineering, technology and waste management (SETWM-20) Johannesburg, South Africa, pp 175–179, 16–17 Nov 2020. ISBN: 978-93-86878-46-5

Paradis E, O'Brien B, Nimmon L, Bandiera G, Martimianakis MA (2016) Design: selection of data collection methods. J Grad Med Educ 8(2):263–264

Piesse M (2016) South Africa: drought threatens food, energy and water security. In: Strategic analysis paper, global food and water crises research programme

Tu Y, Wang R, Zhang Y, Wang J (2018) Progress and expectation of atmospheric water harvesting. Joule 2:1452–1475. https://doi.org/10.1016/j.joule.2018.07.015

The manufacturer's authorised representative in the EU is Springer Nature Customer Service Centre GmbH, Europaplatz 3, 69115 Heidelberg, Germany. If you have any concerns regarding our products, please contact ProductSafety@springernature.com

Printed and bound by CPI Group (UK) Ltd, Croydon, CR0 4YY
30/09/2026
02225315-0001